Using Lasers as Safe Alternatives for Adhesive Bonding:

Emerging Research and Opportunities

Barbara Ewa Ciecińska
Rzeszow University of Technology, Poland

A volume in the Advances in Chemical and Materials Engineering (ACME) Book Series

IGI Global
PUBLISHER of TIMELY KNOWLEDGE

Published in the United States of America by
IGI Global
Engineering Science Reference (an imprint of IGI Global)
701 E. Chocolate Avenue
Hershey PA, USA 17033
Tel: 717-533-8845
Fax: 717-533-8661
E-mail: cust@igi-global.com
Web site: http://www.igi-global.com

Library of Congress Cataloging-in-Publication Data

Names: Ciecińska, Barbara Ewa, 1972- author.
Title: Using lasers as safe alternatives for adhesive bonding : emerging research and opportunities / by Barbara Ewa Ciecińska.
Description: Hershey, PA : Engineering Science Reference, an imprint of IGI Global, [2020] | Includes bibliographical references and index. | Summary: "This book presents various structural materials that can be glued together using adhesives and various types of substances of natural and synthetic origin that are used for adhesive bonding"-- Provided by publisher.
Identifiers: LCCN 2020002720 (print) | LCCN 2020002721 (ebook) | ISBN 9781799846345 (hardcover) | ISBN 9781799853374 (paperback) | ISBN 9781799846352 (ebook)
Subjects: LCSH: Adhesives. | Lasers--Industrial applications.
Classification: LCC TA455.A34 C54 2020 (print) | LCC TA455.A34 (ebook) | DDC 620.1/99--dc23
LC record available at https://lccn.loc.gov/2020002720
LC ebook record available at https://lccn.loc.gov/2020002721

This book is published in the IGI Global book series Advances in Chemical and Materials Engineering (ACME) (ISSN: 2327-5448; eISSN: 2327-5456)

British Cataloguing in Publication Data
A Cataloguing in Publication record for this book is available from the British Library.

All work contributed to this book is new, previously-unpublished material.
The views expressed in this book are those of the authors, but not necessarily of the publisher.

For electronic access to this publication, please contact: eresources@igi-global.com.

Advances in Chemical and Materials Engineering (ACME) Book Series

ISSN:2327-5448
EISSN:2327-5456

Editor-in-Chief: J. Paulo Davim, University of Aveiro, Portugal

MISSION

The cross disciplinary approach of chemical and materials engineering is rapidly growing as it applies to the study of educational, scientific and industrial research activities by solving complex chemical problems using computational techniques and statistical methods.

The **Advances in Chemical and Materials Engineering (ACME) Book Series** provides research on the recent advances throughout computational and statistical methods of analysis and modeling. This series brings together collaboration between chemists, engineers, statisticians, and computer scientists and offers a wealth of knowledge and useful tools to academics, practitioners, and professionals through high quality publications.

COVERAGE

- Nanomaterials
- Fracture Mechanics
- Composites
- Thermo-Chemical Treatments
- Multifuncional and Smart Materials
- Computational methods
- Ceramics and glasses
- Heat Treatments
- Industrial Chemistry
- Biomaterials

IGI Global is currently accepting manuscripts for publication within this series. To submit a proposal for a volume in this series, please contact our Acquisition Editors at Acquisitions@igi-global.com or visit: http://www.igi-global.com/publish/.

The Advances in Chemical and Materials Engineering (ACME) Book Series (ISSN 2327-5448) is published by IGI Global, 701 E. Chocolate Avenue, Hershey, PA 17033-1240, USA, www.igi-global.com. This series is composed of titles available for purchase individually; each title is edited to be contextually exclusive from any other title within the series. For pricing and ordering information please visit http://www.igi-global.com/book-series/advances-chemical-materials-engineering/73687. Postmaster: Send all address changes to above address.

Titles in this Series

For a list of additional titles in this series, please visit:
http://www.igi-global.com/book-series/advances-chemical-materials-engineering/73687

Natural Fiber-Reinforced Geopolymer Composites Synthesis, Properties, and Applications
It-Meng (Jim) Low (Curtin University, Australia) Thamer Alomayri (Umm Al-Qura Univeristy, Saudi Arabia) and Hasan Assaedi (Umm Al-Qura Univeristy, Saui Arabia)
Engineering Science Reference • ©2020 • 230pp • H/C (ISBN: 9781799834595) • US $185.00

Nanotechnologies and Clusters in the Spaces of Higher Dimension Emerging Research and Opportunities
Gennadiy Vladimirovich Zhizhin (Independent Researcher, Russia)
Engineering Science Reference • ©2020 • 240pp • H/C (ISBN: 9781799837848) • US $195.00

Processing, Structure, and Applications of High-Temperature Materials
Brahma Raju Golla (National Institute of Technology Warangal, India)
Engineering Science Reference • ©2020 • 300pp • H/C (ISBN: 9781799831792) • US $195.00

Metaheuristic Approaches for Optimum Design of Reinforced Concrete Structures Emerging Research and Opportunities
Aylin Ece Kayabekir (Istanbul University-Cerrahpaşa, Turkey) Gebrail Bekdaş (Istanbul University-Cerrahpaşa, Turkey) and Sinan Melih Nigdeli (Istanbul University-Cerrahpaşa, Turkey)
Engineering Science Reference • ©2020 • 229pp • H/C (ISBN: 9781799826644) • US $185.00

Properties, Techniques, and Applications of Polyaniline (PANI) Thin Films Emerging Research and Opportunities
Subhash Chander (Indian Institute of Science Education and Research, Mohali, India) and Nirmala Kumari Jangid (Banasthali University, Tonk, India)
Engineering Science Reference • ©2020 • 236pp • H/C (ISBN: 9781522598961) • US $165.00

IGI Global
PUBLISHER of TIMELY KNOWLEDGE

701 East Chocolate Avenue, Hershey, PA 17033, USA
Tel: 717-533-8845 x100 • Fax: 717-533-8661
E-Mail: cust@igi-global.com • www.igi-global.com

Table of Contents

Preface

Adhesive bonding, as a technique of joining machine parts, has been known to man since the beginning of its existence. Already in the earlier centuries various natural substances were used to permanently bond or seal objects. Over the years and during the course of civilization, not only materials but also manufacturing technologies have been improved.

Currently, many different construction materials are glued in the industry using a variety of widely available adhesives.

The first part of the monograph consists of four chapters, which present basic issues related to adhesive bonding technology. The first chapter characterizes the use of gluing technology, its advantages and disadvantages. Types of adhesives, their features, selection criteria, examples of applications are described. The mechanisms of adhesive bond formation are characterized and selected joint constructions are shown.

Chapter 2 describes selected groups of construction materials used in modern industry, how to shape them, and forms available on the market. Attention was paid to the suitability of specific semi-products for gluing structures. In Chapter 3, individual stages of gluing are presented. In addition, requirements for the preparation of the surface to be glued, adhesive application techniques, necessary equipment are described in this chapter. Also in this chapter are described the phenomenon of adhesion inextricably linked to the creation of adhesive joints and methods for determining the so-called surface wettability.

Subsequently, in Chapter 4, much attention was devoted to the description of traditional, widespread ways of implementing the first stage of gluing - surface preparation. Many different variants of this stage have been described, depending on the materials combined and technical possibilities. Attention was paid to various chemical substances, tools and equipment necessary for proper surface preparation before applying the adhesive.

The second part of the monograph is devoted to an issue less frequently appearing in the literature, namely the health and environmental context of technology universal in the industry.

The need to respect the environment, the necessity for sustainable development as well as the respect for natural resources and attention to the quality of life have been presented in various documents around the world. The unprecedented overproduction and environmental burden of municipal, industrial and hazardous waste are occurring on a global scale. The impact of the current way of life can be observed, for example by following media reports on waste problems, as well as by observing alarming phenomena occurring in nature - unprecedented weather and climate problems, soil contamination, pollution of water and atmosphere, disturbance of biological balance in nature, the increasing problem of human health loss and the occurrence of civilization diseases on a large scale.

In connection with the issues outlined above, it is important to realize where the source of these problems is. Are technological operations indifferent to human health and nature? Unfortunately, the answer is not always positive, which can be exemplified by gluing.

Therefore, it is necessary to take appopriate measures to restore the balance in nature and improve living conditions, also for future generations.

In connection with the above issues, it should be realized that technological operations are not always indifferent to human health and nature. Better, safer manufacturing techniques are sought.

Chapter 5 discusses modern machining tools - lasers. The chapter presents the specification of laser operations as well as the types and characteristics of laser radiation. Selected areas of laser applications in modern industry and their advantages are also described in this chapter. The thesis was put forward that laser surface processing can replace the existing methods of treatment which have a harmful effect on human health and are burdensome for the environment.

Chapter 6 shows the results of the author's research demonstrating that it is possible to remove contaminants from the surface, as well as to create favorable adhesion conditions, necessary in bonding. The analysis of surfaces prepared by laser of the selected materials is presented, both in terms of specific and mechanical adhesion. The results of tests of static shear strength of joints made according to the proposed changes are also presented. The obtained results raise hope that a wider choice of lasers will be applied in assembly technology.

The presented possibility of using laser in adhesive technology is important for the reasons presented in Chapter 7. It describes a number of health and environmental threatening situations in adhesive operations carried out conventionally with comparison to the technology using lasers. The complexity of the analyzed problem is visible especially at the stage of surface preparation for bonding. A human operator comes into contact with many toxic chemicals and is exposed to diseases. In addition, there are other environmental aspects, such as air pollution, water and energy consumption, danger due to use and storage of chemicals. The environmental problem of the analyzed technological operations is also visible after the manufacturing process is completed.

As a result of adhesive bonding operations a huge amount of hazardous waste is produced, these include used packages of adhesives, hardeners, machining and measuring fluids in metal or plastic containers (sometimes under pressure), worn instruments and tools. Contaminated solutions of acids, alkalis, salts, especially those used for baths during pickling, anodizing and others pose a big threat to the environment. Potential risks to human health and the environment can occur during overflow, tank maneuvering and storage in huge containers. In addition, caution and care must be exercised during storage and transport when there is a risk of spillage, self-ignition, fire, explosion, but also escalation of events in the form of domino effect and as a result of significant losses in the health and life of people, property, and above all, components of the natural environment.

In recent years, a significant increase in the popularity of lasers in the industry has been observed, as presented in Chapter 8. The reason for this phenomenon, in addition to the needs and production capabilities, such as accuracy, repeatability, durability, speed of processing, flexibility in changing the operating parameters of the laser head, easy access to construction materials, is also the environmental aspect. As a result of laser beam processing, often controlled by a computer or coupled with industrial robots, fewer products that are not in conformity with requirements are produced, no production "to stock" is necessary, and finally there is a noticeable lack of waste characteristic of conventional processing. Lasers have found a special place in modern plants, in some countries, still few, and sometimes only pilot plants, built and managed according to the concept of Industry 4.0. So far, these plants, in most cases, offer sheet metal products that can be easily cut, bent, welded, marked, etc.

Unfortunately, lasers for cleaning elements are less common, although contamination has already been removed from various constructions, from casting and injection moulds, bridges, to aircraft fuselages, unique museum objects (sculptures, paintings) and archaeological discoveries.

Why not, then, consider using laser to clean the surface before bonding?

After the analysis, carried out by the author on the basis of her own experimental research, it can be concluded that there is a possibility to eliminate, first of all, a considerable amount of poisonous chemicals, which pose a hazard to people and the environment, as well as the load of liquid and solid waste. It should be remembered, however, that the technique of laser processing is not completely free from defects in the context of the problems analyzed in the monograph. Unfortunately, it is not possible to avoid all the inconveniences that are residues of traditional processing. It is still necessary to apply certain working procedures in order to ensure that operations with lasers are performed safely.

Thinking in perspective, also in the context of changing legislation, production guidelines (the search for the best available BAT technologies, implementation of environmental directives), the continuous technical development, the search for other technological solutions should be evaluated positively. The research findings of many scientists (including the author's experience in cooperation with the departments in Poland, Slovakia and United Kingdom), give reason to hope that it will be possible to implement the technique of laser bonding without the use of adhesives.

To sum up, the development of civilization and technology we are witnessing on an unprecedented scale is undoubtedly the source of satisfaction linked to a number of improvements and inventions. It is important to be aware, however, that intense production, consumption of raw materials, their non-renewable nature and the resulting waste may be the problems a contemporary man has to face. Therefore, by following the principles of sustainable development, there remains nothing else but to seek technologies that are people-friendly and environmentally friendly.

Acknowledgment

The publishing of a monograph on lasers and the possibility of their application in adhesive bonding operations has been my dream for several years. This work was completed with the spiritual and material support of many people and institutions.

Of the many who have given me help and encouragement in the preparation of this book, I am especially indebted to my Parents and my closest family. This monograph would not be a reality today without their cooperation and moral support for many years.

I am also very grateful to my current mentor Professor Tomasz Więcek, the President of Science for Industry and Environment Foundation, who with great determination and kindness, gave me a lot of encouragement and guidance in my project and research work.

Special thanks, also, go to my colleague Professor Lydia Sobotova from Technical University of Kosice (Slovakia), from whom I took inspiration to work on the use of lasers in adhesive bonding technology. Also, I thank my colleagues from Rzeszow University of Technology (Poland) and University of Rzeszow (Poland) for their kindness and help in research work.

In addition, I thank my dear colleagues Katarzyna Kurek and Iwona Jagusztyn who worked hard and suggested numerous language improvements.

A debt of gratitude is also due to former employees of the Monuments Conservation Center in Poznań (Poland), as well as colleagues from TRUMPF Poland (especially to the President Andrzej Janiszewski and Mr Ireneusz Gągała), and SPI Lasers United Kingdom - I would not have known the fantastic world of lasers without you.

Barbara Ewa Ciecińska

Chapter 1
Adhesive Bonding:
Types of Adhesives and the Preparation of an Adhesive Mixture

ABSTRACT

This chapter presents the application of adhesive bonding technology with a long tradition of use in manufacturing processes. It has an established position among a number of different ways of permanently joining elements together. This work offers a list of many advantages of this technology, alongside the difficulties and limitations in the use of adhesive bonds. The basic concepts of adhesives and their components are described, and the terms related to the bonding process are presented. Attention was paid to the great variety of adhesives and many different criteria of their division and selection to a particular technological situation. In the following chapter, examples of the application of different types of adhesives with regard to using materials and constructions can be found as well as methods and requirements for the preparation of the finished adhesive. Mechanisms of the formation of an adhesive bond are characterized, and various constructions of adhesive bonds that are necessary to form a durable bond of elements are presented.

DOI: 10.4018/978-1-7998-4634-5.ch001

APPLICATION OF ADHESIVE BONDS

Currently, adhesive bonding has a strengthened presence in the industry as a technique of connecting elements together. In the production of a wide variety of products, there is a large number of components that are permanently bonded together in this way. It is not possible to separate the bonded parts without destroying them, as opposed to the equally commonly used separable connections, e.g. twisting (Sobieszczański J., 2004).

However, despite some inconveniences, adhesive bonds are used in almost all industries. One of such industries is the furniture industry, where furniture structures, upholstery and decorative elements are bonded together. Adhesive bonding connects elements of various types of metal tanks, packages of transformer plates. In addition, adhesive bonding is increasingly replacing classic methods of bonding used in machine and vehicle construction, for example, to connect parts of bodies made of steel or aluminum plates (where adhesion takes place in welding or soldering, especially in hard-to-access places or elements vulnerable to high temperature deformality) (Sobieszczański J., 2004).

In automotive industry, entire car bodies are adhesively bonded together - the first Ford Taunus construction in 1960 may be a distinct example of it. In this first Ford construction, engine and boot covers, fenders and a folding roof were bonded together with synthetic rubber adhesive. In the automotive industry, sealing pastes made of polyvinyl chloride are also used. For example, in 1949 Chrysler replaced riveting of brake linings with adhesive bonding using phenol-formaldehyde resin (Kniewald D., 1994).

Adhesive bonding has become a particularly useful technology in aviation and cosmonautics as it is possible to manufacture ultra-lightweight, yet very durable structures. Reduction of aircraft weight by 1 kg leads to savings of about 20 to 40 thousand dollars. Covering the outer jacket of space shuttles (e.g. Columbia) with 30 000 bonded silicon wafers as a thermal shield, resistant to temperatures of 1500°C is a well known solution. Another characteristic structure among flying objects is the so-called honeycomb structure bonded in, among others, rotor blades of helicopters Mi 2 (Cagle Ch. V., 1977).

In the machine industry, in assembly operations, adhesive bonding makes it possible to eliminate backlash, fill gaps, ensure tightness of the structure, protect against loosening and vibrations in screw, pin, spline and hub-shaft connections. Thus, it is possible to extend the life of the product and reduce the costs associated with labor-intensive processing (Puff T. & Sołtys W., 1980).

Adhesive bonding is also used to repair damaged machine elements, for example to fill in short run castings or cracks in cast machine bodies and to regenerate worn surfaces working in sliding connections. For example, when an adhesive layer is applied to a worn journal, the adhesive layer is turned to the desired journal dimension after curing (Sobieszczański J., 2004). As a structural repair technique, bonding is also used in military applications as it is easy to perform operations with adhesives under field conditions. Examples are resin filling in pump bodies, sealing of radiators, motor vehicle engines caused by intensive exploitation or the effect of using arms. In combat conditions, metal tanks, rubber elements of tires and aluminum linings are bonded together. It is common practice to reproduce the damaged structural elements using adhesives, which is useful especially in situations where it is not possible to perform a professional repair. In this way, the replacement element is used on an ad hoc basis and the device recovers its function. To repair combat equipment, the so-called adhesive composites are used, i.e. resins with fillers, e.g. in the form of textiles, foams, metal cores. A good application example are BELZONA materials used by armies of different countries in NATO (Godzimirski J. & Others, 2010).

Unusual applications of adhesive bonding operations include bonding in medicine not only to manufacture laboratory equipment, but also to tend cuts. The surgery of adhesive bonding of the severed body enables faster closure of traumatic and surgical wounds compared to traditional sewing. Additionally, cosmetic appearance of the scar is similar to the one treated in a traditional way. The first adhesives used with tissues turned out to be toxic, however further research in the first half of the 1970s resulted in non-toxic fibrin binders used to bond cuts, nerves, trachea, bronchi and bone adhesives. Another example of bonding applied in medicine is the UV-curing adhesive used in dentistry, invented in the 1990s (Godzimirski J. & Others, 1997).

At the current stage of development of adhesive technology, there are many possibilities to select an appropriate adhesive for suitable production or application requirements. There are adhesives known to be cured only in water, humid air environment or in atmosphere free of any traces of water vapor. There are also adhesives that cure in the presence of oxygen or in oxygen-free presence. In addition, some adhesives are suitable for joining steel elements operated at elevated temperatures and others at low temperature (Czaplicki J. & Others, 1987). The variety of adhesives makes it possible to successfully apply adhesive technology in many manufacturing processes.

ADVANTAGES AND DISADVANTAGES OF BONDING

Adhesive bonding is a technique of connecting materials together. In principle, it can be said that the available adhesives allow to combine all solids with each other, i.e. metals with metals, but also with rubber, plastics, wood, ceramics, glass, etc. There are also hybrid connections used between different metals: aluminum steel, titanium alloy steel, plastic with wood, etc. (Czaplicki J. & Others, 1987).

In many cases, bonding is characterized by features not possible to obtain with other methods. At the same time, a lightweight, durable, rigid connection is obtained in an economical and efficient way. It should be remembered, however, that adhesives and glue joints are characterized by undoubted advantages, yet also have certain disadvantages and limitations (Cagle Ch. V., 1977).

From among many advantages of bonding, the following may be distinguished: (Godzimirski J. & Others, 2010; Brockmann W. & Others, 2009; Sobieszczański J., 2004; Rośkowicz M., 2013; Cagle Ch. V., 1977):

1. Possibility to combine materials differing in structure and properties (e.g. in the case of metals, an important feature is the possibility to combine materials with a different melting point).
2. Possibility of simultaneous use of a bond connection with another, e.g. mechanical (threaded) connection in order to improve operational properties.
3. Good ratio of strength to weight of the joint (strong ultra-lightweight constructions are possible).
4. The connection does not require any additional fasteners (rivets, screws), there is no weakening of the structure with the holes made for them, there are no protruding elements in the structure (rivet heads, affecting, e.g. aerodynamic properties).
5. Residual stresses are practically non-existent - adhesives are characterized by low shrinkage and high susceptibility in the bonding phase.
6. In general, bonding is carried out cold, i.e. at ambient temperature, and there are known methods of bonding in warm or hot temperatures. However, adhesive bonding does not change the internal structure of the material, unlike welding or soldering. This allows to bond materials that have been previously reinforced, e.g. by crushing (when crushing allows to create a specific state of stress, as well as a harder surface layer and

leave a softer core) or heat treated. The adhesive temperature does not destroy the previously achieved results.

7. Both the susceptibility of the adhesive and the lack of interference in the structure of the material due to temperature, allow the possibility of gluing in the last technological operation of the already finished elements (even after thorough processing).
8. It is possible to modify the composition of the adhesive in order to obtain the desired properties, e.g. to increase its elasticity (yet, sometimes at the expense of other mechanical properties), to improve its electrical and thermal conductivity (e.g. recently applied nano fillers or to reduce them (Ciecińska B. & Kubit A., 2017).
9. The adhesive may be used not only as a joint but also as a protective layer, e.g. against corrosion, or as a sealant.
10. The adhesive bond shows the ability to dam vibrations.
11. There is practically no limit to the thickness of the parts to be joined - thick and thin parts can be successfully joined together.
12. It is possible to bond elements of considerable dimensions as well as small, simple or complicated shape.
13. The gluing procedure is not particularly demanding due to production procedures.
14. The cost of making an aesthetic connection is relatively low.

In many cases, adhesive technology has advantages over soldering, welding or mechanical bonding. However, this does not mean that bonding can replace conventional bonding techniques in all current applications. The implementation of this technique depends on the selected elements or construction and is used if the operational requirements are met and when it is economically justified.

Despite the undoubted advantages, the disadvantages of adhesive bonding are noticeable (Sobieszczański J., 2004; Puff T. & Sołtys W., 1980):

1. Usually the strength properties of glued joints are quite low. Shear strength Rt, depending on the type of adhesive used in machine building, is usually between 10 and 40 MPa and there is quite a significant dispersion of strength properties of joints formed according to the same procedures.
2. Adhesive bonds have low peel strength.
3. Comparing chemosetting and thermosetting adhesives, it can be concluded that a greater dispersion of strength properties occurs with chemosetting adhesives; these adhesives are not resistant to elevated temperatures. It can

be assumed that each increase in temperature above ambient temperature is matched by a certain decrease in strength properties. At 100°C, this reduction in properties may be about 50%, although adhesives resistant to the so-called thermal shocks, i.e. short term increases in temperature up to several hundred Celsius degrees, are already being produced.

4. Adhesives are subject to ageing, which leads to the decrease of strength properties of the joint over the course of time. In addition, adhesives have low creep resistance, i.e. under constant load, permanent deformation increases after a certain period of time. They also show low resistance to moisture and limited resistance to chemicals.
5. Adhesives and materials used for adhesive bonding can be harmful to people and the environment. It is necessary to ensure that safety procedures both during adhesive bonding and waste handling are followed.
6. During installation, it may be difficult to identify non-bonded areas, which may result in the loss of quality.
7. Adhesive bonds are 'sensitive' to the thickness of the adhesive (the rule 'the more adhesive, the better' does not apply here). The best indications in terms of strength are obtained with an adhesive layer with a thickness of 0.05 to 0.1 mm. Going beyond this range may result in a significant decrease in strength properties.
8. In technological processes after bonding, no further processing is planned due to the possibility of damage to the weld.
9. In technological processes: the long curing time is sometimes necessary, which prolongs the production cycle, as well as a significant number of recommended control operations.

FUNDAMENTAL CONCEPTS RELATED TO ADHESIVE BONDING

In adhesive bonding operations adhesive is the basic material applied, alongside connected elements.

An adhesive is a non-metallic substance of organic or inorganic origin which permanently bonds solid materials without causing significant changes in their shape and properties, is inserted between surfaces of adjacent objects and enables their permanent joining. Adhesives belong to the group of surfactants, which also includes paints, varnishes and detergents (Cagle Ch. V., 1977; Czaplicki J. & Others, 1987).

A connection of two or more structural elements, called **an adhesive bond,** is determined by the adhesion of the adhesive to the surface which it is applied to and its internal cohesion. The hardened adhesive, then called a **bond** (preferably to match the strength of the bonded materials) depends on the value of the intermolecular forces (Godzimirski J. & Others, 1997).

The main component of the adhesive is the **binder**; this substance determines the required adhesion to the bonded surfaces as well as appropriate mechanical strength of the joint after its solidification. In many cases the binder is not the only component of the adhesive. Various substances are added to the binder to modify its functional and strength properties. This mixture of the binder and additives is called **adhesive mass**.

The adhesive mass may contain (Czaplicki J. & Others, 1987; Porębska M. & Skorupa A., 1997):

- **a solvent,** i.e. a liquid substance of main organic origin, which reduces the density and viscosity of the binder by dissolving it,
- **a thinner** - liquid used to reduce viscosity and adhesive concentration or to change adhesive mass dispersion; the use of the thinner reduces costs associated with the use of expensive solvents and leads to proper regulation of the evaporation time of volatile adhesive components,
- **a softener** (**plasticizer**) - a substance increasing the plasticity and elasticity of the adhesive joint; softener is used only when it does not have the ability to migrate to the bonded surface from the joint veneer and does not affect it chemically or physically, causing a deterioration in adhesion; otherwise, the use of softener may lead to a decrease in the joint strength,
- **an extender** - a non-volatile component, whose use allows to reduce the use of a more expensive binder (binders); it is usually a solid substance without adhesive properties, insoluble in the adhesive, used to modify its physical properties within a specific use range;
- **a filler** - a non-volatile auxiliary substance with poor adhesive properties; it is mainly used to reduce the production costs of adhesives,
- **a pigment** - an insoluble, opaque, colored powder, usually of mineral origin, used for coloring by direct mixing with adhesive mass. Inorganic pigments give plastics and adhesives, in addition to color, a certain resistance to chemical, thermal and weather conditions. The most common ones are metal oxides or metal sulphides.

A durable adhesive bond between structural plastic components is achieved by using chemically cured adhesives. In order to initiate and accelerate the curing process, a substance or mixture of several substances is added to the adhesive, respectively. Among them, the **hardener** occupies the main position and, when added to the adhesive, causes a chemical reaction of curing in it and takes an active part in the process. The hardener should be distinguished from a catalyst, which, by accelerating the reaction, is not part of the polyreaction product and is not subjected to any transformations.

In some cases, **accelerators**, i.e. compounds accelerating catalytic curing of glue, as well as **activators** or **primers** - substances which, by chemical interaction with other compounds, initiate their specific chemical reactions. **Inhibitors** are chemicals that are added to adhesives to increase their lifespan.

The life of an adhesive is the time from the moment of its production until the occurrence of symptoms of deterioration of its functional properties, resulting from physical or chemical changes, for example gelling, delamination. For the user, the durability of an adhesive is usually measured by its warranty period.

Adhesive bonding is the transition from the plastic state of an adhesive in a joint to the solid state. It may be of a physical nature, e.g. by evaporation of the solvent or solidification of the molten adhesive; or of a chemical nature, e.g. by polycondensation, polymerisation or polyaddition. The bonding process can take place at low or elevated temperatures, hence the terms **cold** or **hot bonding** were coined. In the case of adhesive bonding due to solvent evaporation, the following terms are used: drying time open (open time) and drying time closed.

Open time (**open drying time**) - otherwise known as **drying** - is the time measured from the moment the adhesive is applied to the surfaces to be bonded. During this time, the adhesive coating achieves optimum adhesion by evaporating a large amount of solvent.

The closed time (**the drying time of a closed** adhesive joint) is the period from the bonding of the adhesive coated elements dried with the adhesive to the complete drying of the joint due to the evaporation of the solvent residue.

The use of adhesive bonding and appropriate conditions, especially appropriate temperature, affects the so-called **curing time**, i.e. the period from the assembly coated with the adhesive surface to the achievement of optimal strength by the adhesive joint. Curing is usually carried out under increased pressure, called **pressing pressure** or **bonding pressure**. It depends on the pressure whether the thickness of the veneer will reach the optimum value, influencing the strength of the joint.

In addition to the above mentioned fundamental terms for adhesives, there are other terms, also related to the process of adhesive bonding.

Duration of adhesive life - workability time of adhesive mass used to bond the surface, understood as the time from the moment of preparing the mass until it begins to lose its properties.

The curing cycle - the period from the addition of the hardener to the resin to the completion of the curing process, including all additional operations, e.g. heating or pressing necessary to obtain the hardened product.

The seasoning time of adhesive joints - the period from the completion of the proper process of gluing to achieving maximum mechanical strength of the joint.

Bondability - the ability of adhesives to stick to various surfaces during drying; after complete curing, gelling or drying, the adhesiveness disappears.

In the context of creating an adhesive bond, the following terms may also be found:

Failure to cure - inadequate cure (or incomplete cure) that can be recognized by non-stickiness and low strength of the solidified adhesive.

Resistance to ageing - resistance to the influence of the surrounding environment, oxygen contained in air, elevated temperature, light, moisture or similar factors.

TYPES OF ADHESIVES

The characteristic features of adhesives are one of the most important factors influencing the properties of the bonded joint. There are many requirements towards adhesives, e.g. requirements concerning the strength and performance of the joint as well as technical requirements. When constructing bonded joints, the strength characteristics of the joint must be taken into account. Thus, if the shear strength of the bonded elements is similar to that of a homogeneous element, such adhesives can be evaluated positively. However, at the same time, conditions and operating life of the bonded component need to be recognized (Puff T. & Sołtys W., 1980).

The basic criterion for selecting an appropriate adhesive is the material to be bonded. In this regard, the following adhesives listed in Table 1 are distinguished (Cagle Ch. V., 1977; Godzimirski J. & Others, 1997).

From the point of view of structural requirements, adhesives can be divided according to shear strength. For example, high shear strength adhesives can be used to join elements under shear stress, low strength adhesives can be used

Table 1. Adhesive types according to application

Adhesives due to the type of the material to be bonded
For metals For wood For plastics/rubber/carpets/tissue floor coverings For wallpaper/paper/wall coverings For glass

Table 2. Division of adhesives according to joint strength

Type of adhesive	Strength properties
Pressure sensitive adhesives	Low strength
Mounting adhesives	Medium strength
Structural adhesives	High-strength

to join elements without stress (Table 2) (Cagle Ch. V., 1977; Godzimirski J. & Others, 1997).

Depending on the origin, adhesives can be divided into natural (e.g. starch, cellulose) and synthetic (e.g. synthetic) adhesives. Most of the current adhesives are synthetic. Adhesives of both types are presented in the table below (Table 3) (Porębska M. & Skorupa A., 1997).

Adhesives may bind by physical processes mainly due to evaporation of one of the components, e.g. water or solvent, or due to chemical reactions. These processes may have different varieties (Table 4) (Cagle Ch. V., 1977).

Table 3. Division of adhesives according to their origin

Natural adhesives	Synthetic adhesives	
Starch Glutinous Casein Cellulose derivatives Collagen Dextrin	Amino Phenolic Reserve Epoxy Polyurethane Polyvinyl acetate Vinyl acrylic	Acrylic Based on polyvinyl chloride Rubber Cyanoacrylate Polyester Anaerobic Methacrylate

Table 4. Division of adhesives according to the joint bonding mechanism

Physical process	Chemical process
Solvent evaporation Evaporation of water Melted adhesive build-up	Polymerization Latticing

The division is often used because of the binding agent. This is when self-curing, chemically-curing and thermosetting adhesives are distinguished. Chemosetting adhesives bind under the influence of a hardener, usually at room temperature. In the case of thermosetting adhesives, the binding is caused by an increase in their temperature above 100°C. Adhesives are manufactured in liquid or solid form. In solid form they need to be dissolved or melted (Table 5) (Porębska M. & Skorupa A., 1997).

Table 5. Division of adhesives according to their form

Liquid adhesives	Solid adhesives
Solvent-based Experimental Aqueous solutions Liquid resins Polymerizing monomers	Solid resins Adhesive foils Adhesive films Powders and granules to dissolve

In addition, adhesives can be single or multi-component types. Single-component adhesives are commercially available as ready-to-use compositions, or only requiring dissolution, dilution or addition of filler. In the case of multi-component adhesives, the combination of components takes place immediately before the application of the adhesive layer (Sobieszczański J., 2004). Two-component adhesives are common - the adhesive is delivered to the recipient in the form of two separately packed components, which must be mixed immediately before use. They may be portioned in separate containers or in special double cartouches. One component is based on synthetic resin or rubber, the other on hardener, crosslinking agent or vulcanizing agent.

From the point of view of technological requirements, a distinction can be made between cold cured, warm cured and hot cured adhesives (Table 6) (Porębska M. & Skorupa A., 1997; Ebnesajjad S., 2008).

Table 6. Division of adhesives according to curing conditions

Types of adhesives	Curing conditions
Cold hardened	At a temperature of 20-30°C
Heat hardened	At temperatures of more than 30°C to approximately 99°C
Hot hardened	At temperatures above 100°C

Each of these groups may contain adhesives that are cured with or without pressure. The curing process can also be carried out in two ways, as shown in the table below (Table 7) (Ebnesajjad S., 2008).

Table 7. Division of adhesives according to curing process steps

Method of hardening	Characteristics
Single-step curing adhesives	The process is carried out without changing the hardening conditions.
Two-stage hardened adhesives	After the precure, usually cold, is followed by a process called postcure, warm or hot, at a temperature specified by the manufacturer.

Due to the fact that at present adhesives with synthetic and modified resins are preferred, the division into two basic groups is applied in practice: thermoplastic adhesives and thermosetting and chemosetting adhesives (Puff T. & Sołtys W., 1980).

The group of thermoplastic adhesives includes adhesives used to bond metals to other materials, which are characterized by their ability to become liquid under the influence of the temperature of the solvent. It is a reversible process as the resin hardens if the heat or chemical agent is eliminated. These adhesives consist of thermoplastic resin and modifying additives such as plasticizers and adhesion promoters. Thermoplastic resins belong to the group of plastics of relatively low strength and are usually characterized by low hardness and heat resistance (it varies from 40 to 80°C). This group includes polyvinyl chloride based adhesives and polyacrylic adhesives (Cagle Ch. V., 1977).

The group of thermo and chemically cured adhesives include phenolic, two-component epoxy, polyurethane, silicone and other adhesives characterized by high mechanical strength and resistance to weather conditions. A special attention should be paid to epoxy adhesives due to their relatively high shear strength of approx. 15 MPa when cold gluing and approx. 40 MPa when hot gluing. These are used for bonding steel, aluminum, thermosetting plastics, glass and porcelain.

Examples of adhesive applications are given in the table below (Table 8) (Ebnesajjad S., 2008).

The examples of adhesives there are in the Figure 1.

Table 8. Selected types of adhesives and their applications

Type of adhesive	Bonded structures
Phenolic and phenol-formaldehyde resins	Load-bearing elements for aluminum and aluminum alloy aircraft, for wood, cardboard, etc.
Polyamide adhesives	Metal elements of aircraft structures
Epoxy resins	Steel structures, lightweight metals and their alloys, plastics (with exceptions), laminates, glass and ceramics
Polyurethane adhesives	Rubber, metals, leather, wood, plastics (even those difficult to be bonded together)
Cyanoacrylate adhesives	For fast bonding of plastics, textiles, metals, wood and textiles
Methacrylate adhesives	For structures exposed to water and chemicals (except for solvents)
Rubber adhesives	E.g. in the automotive and furniture industries to bond leather, textile and rubber
Casein adhesives	For structures made from wood, cardboard, leather, paper
Phenolic adhesives	Structures exposed to mould, bacteria, water and acids, for joining parts made from metals and plastics
Anaerobic adhesives	For sealing threads, flanges, securing threads against automatic unscrewing

Figure 1. Examples of adhesives: a) one-component, b) two-component, c) self-adhesive

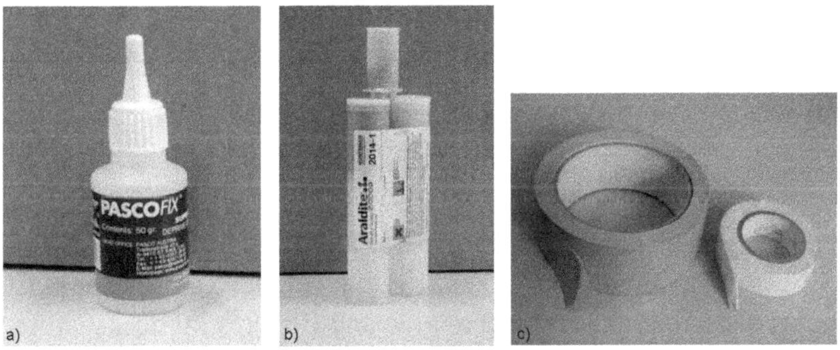

In practice, the basic characteristics of the adhesive are given, including (Habenicht G. 2009):

- lifetime of the adhesive,
- viscosity,
- fragrance,
- chemical type,
- color,

- mixing ratios in the case of compound adhesives,
- modulus of longitudinal elasticity,
- Poisson's coefficients,
- curing time,
- range of applicability,
- other, depending on the manufacturer.

Examples of commercial information for selected adhesives are given in the Table 9.

Table 9. Specification of selected adhesives (Farnell, 2019)

Feature	Trade name		
	Loctite 3422	Araldite 2014-1	Araldite 2021
Intended use	For metals	For metals	For metals
Numerous ingredients	2	2	2
Color after mixing	Light yellow or white	Gray	White
Specific gravity	1,2 g/cm3	1,6 g/cm3	1,03 g/cm3
Mixing ratios	1:1	2:1	1:1
Open time of glue	3-4 min. at 25°°C	60 min. at 23°°C	3-4 min. at 23°°C
Curing time	120 min. at 25°°C	180 min. at 23°°C	8-12 min. at 23°°C
Curing temperature	25°C	23°C	23°C
Viscosity	Slightly thixotropic*	Thixotropic*	Thixotropic*
Modulus of elasticity	1298 N/mm2	4000 N/mm2	2300 N/mm2
Chemical type	Epoxy	Epoxy	Methacrylate

*Thixotropy – the so-called "liquid memory" property of some liquids, in which viscosity is dependent on shear forces and manifests itself through, among other things, a decrease in viscosity during mixing; thixotropic adhesives are easy to mix and do not drip when applied.

A specific type of adhesive is gap-filling adhesive, which does not contain volatile solvents or contains only a small amount of them, and can be plastically deformed (formed). It can be used to glue and fill gaps in the structure at the same time. Gap-filling adhesive is not only a sealing compound because no strength requirements are imposed; in this case the terms "putty" or "sealant" are used.

PREPARATION OF THE ADHESIVE MASS

Commercial adhesives have different forms regarding state of aggregation and quantity of components. One-component adhesives are ready to use immediately, solid adhesives are removed from the container, liquid adhesives are squeezed out of the tube, cartridge or poured out. In the case of multi-component adhesives, it is necessary to mix the components properly in order to obtain a mass with the desired properties.

Two-component adhesives are, in most cases available in pre-prepared double cartridges containing a separate adhesive and a second component, such as a hardener, activator, reaction catalyst and others. Cartridge manufacturers indicate the mixing ratio of the ingredients and special mixing tips are used to obtain a suitable mass. When squeezing substances using a special tool called "gun", components pass through a suitably shaped system of channels, mixing themselves to obtain a uniform mass (Figure 2 - 4). Such a method is easy to use in practice, the extrusion of the adhesive with the "gun" can be done manually or with the use of guns or pneumatic nozzles (suited to

Figure 2. The full set: resin and hardener in one cartridge with an appropriate pressing insert used for squeezing - from left: 1:1 and 2:1

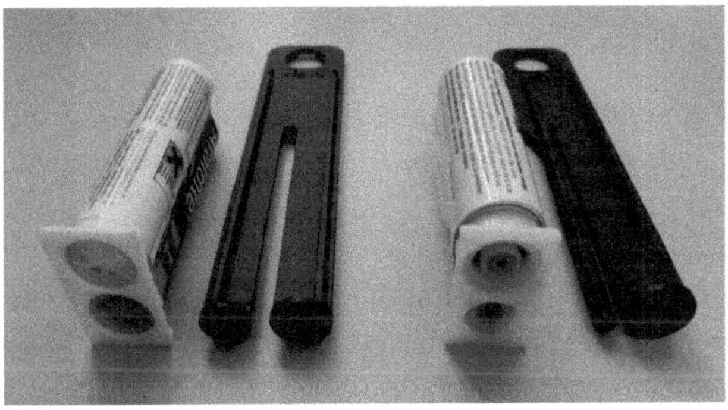

Figure 3. Adhesives in different packages: large 200ml from the top, and small 50ml

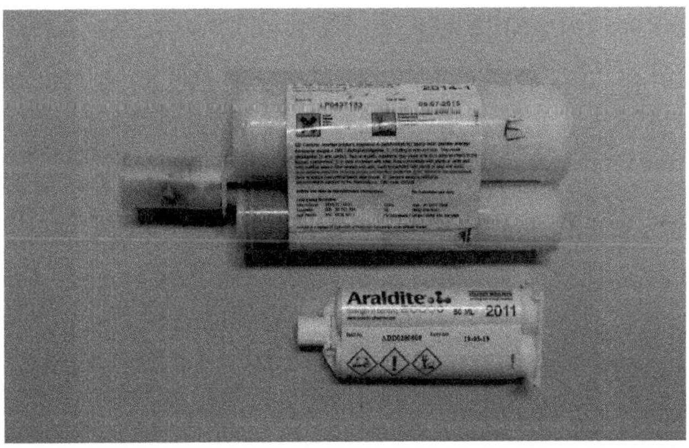

Figure 4. Instruments for simultaneous mixing and pressing of the adhesive: a) for large packages,(b) for small packages, (c) mixing tips

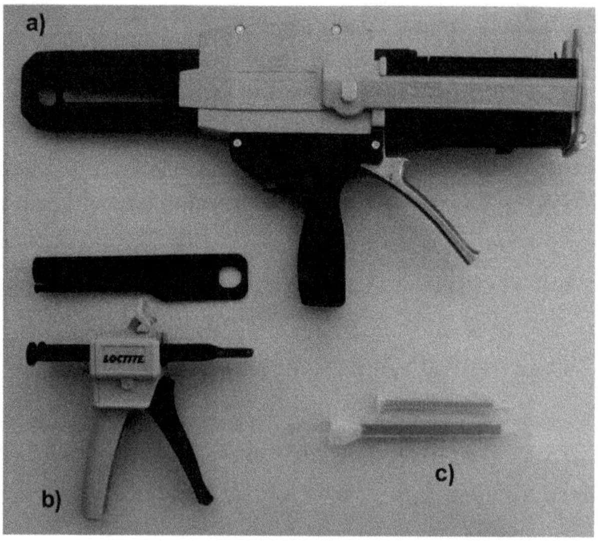

mechanization and automation in the production process), most often with simultaneous heating of the adhesive. In addition, obtaining a suitable adhesive mixture is always guaranteed.

A little more attention needs to be paid to adhesives whose components are delivered separately. Measuring and mixing of the ingredients requires attention and precision. In addition, specific environmental conditions under which the mass is prepared due to the specificity of the chemical reactions must be ensured. The following may affect the final effect:

- **proportions** – inappropriate ones can lead to poor mass quality, for example, both too small or too big quantity of hardener leads to incomplete cross-linking of the polymer (Figure 5). Then, the adhesive joint remains soft, ductile, sticky and not strong enough (Czaplicki J. & Others, 1987).

At the same time, by mixing the components in appropriate proportions, properties of the joint can be controlled. Epoxy adhesives are a good example of this, characterized by the so-called epoxy number *le*, which is used to calculate the amount of the hardener. For DEH24-DER (tri-ethylene diamine) hardener, the stoichiometric amount of the hardener is calculated from the formula (Czaplicki J. & Others, 1987):

Figure 5. Epoxy resins (a, b) and hardeners (c, d) delivered separately

$X = 24.3\, le,$

where:

X is the number of grams of hardener per 100 g of resin
le is epoxy resin number in gram equivalents per 100 g of resin

In the case of this hardener, it is not recommended to exceed the amount by ±5%, as its excess results in leaving the not fully bonded hardener in the mass, whereas its deficiency causes the reduction of reactivity and the reduction of, among others, mechanical properties, heat resistance of the joint and chemical resistance.

In the case of other hardeners, on the other hand, there are manufacturer's recommendations for controlling the properties of the joint. For instance, VERSAMID 125 (polyamide C) hardener, where, for example, with ARALDIT-F resin, different proportions can be used (in weight parts, resin: hardener): 70:30, 60:40, 50:50 and 40:60. Masses a bigger amount of the hardener enable to obtain a more flexible, impact-resistant, yet less hard and less heat-resistant joint.

- **presence of water or carbon dioxide in the air** - there are adhesives or additional substances (especially hardeners) sensitive to high humidity or other contaminants. In this case, water or CO_2 molecules are absorbed and cause polyreaction, which deteriorates strength properties,
- **temperature** - there are hardeners which, when mixed at room temperature, do not generate a complete cure. The mixture of the adhesive composition is then heated taking care of the proportions. Additional heating can also be used for very dense adhesives, which cannot be diluted. In this case, a slight heating improves wettability, reduces viscosity and facilitates adhesive application. The need for heating can also be determined by the form of adhesive - adhesives in solid form, as rods, powders, membranes soften when the melting point is exceeded and only in this state can they be applied. In practice, a reverse solution is also found when rubbing a heated surface with an agglomerated adhesive. Powdered adhesive can be spread on a cold surface and then heated using a dryer or a radiator until a liquid mass is obtained,
- **mixing accuracy of the components** – enables to obtain a homogeneous mass with specific properties. In the case of liquid substances delivered in tubes, this problem is solved by the construction of the tip. However, in the case of glues and additives added separately, it is important to mix using mixers until a homogenous consistency is obtained, ensuring that no air bubbles get into the mixture. They can impair the mechanical properties of the bond, causing the creation of cavities between surfaces (filled with gas, not adhesive).

The problem of mixing and obtaining the appropriate particle distribution is particularly important when adding solid particles to the mass. In recent years, the so-called nanofillers, i.e. particles having at least one dimension, whose natural unit of measure is the nanometer (Kelsall R.W. & Others, 2012), have enjoyed wide interest. As evidenced by experimental studies, even small amounts of nanofillers resins should be added to the adhesive. The examples were given in the paper by (Marcq F & Others, 2011), i.e. nanotube additives may be used in electronics due to better thermal conductivity (an important feature due to heat dissipation in electronic systems). The addition of 5% of carbon nanotubes allows to improve thermal conductivity by about 10%. Also, physical properties may change – the glass transition temperature increases by approx. 20% after the addition of 5% of Al_2O_3 nanotubes and carbon nanotubes by over 16% (Ciecińska B. & Kubit A., 2017). Mechanical properties of finished joints also change: static peel strength increases by over 15% (with the addition of SiO_2 and Al_2O_3 nanopowders) and fatigue strength increases by over 12% (Zielecki W., Kubit A., 2013). Due to positive effects, it is important to ensure that the properties of the adhesive layer remain uniform. To achieve the desired effect, the resin and the filler are mixed directly with the nanofiller in high-speed agitators, rolling mills or extruders. During such mixing, the viscosity of some polymers increases rapidly after adding a bigger amount of nanofiller. To improve wettability, modifiers (surfactants) are added. Another way is to mix the nanoparticles in a solution of polymer and solvent, after which the solvent is evaporated. When nanoparticles form difficult-to-mix agglomerates, ultrasound of certain amplitude are applied to break them down. Sometimes both methods - direct and solvent mixing - are combined for a better effect (Kelsall R.W. & Others, 2012). If there is a problem of uneven mixing or air bubbles, manual mixing is not recommended, however mechanical flow mixers are used.

In the case of adhesive coatings production, the method of spraying and drying is used to obtain a proper adhesive thickness (Kelsall R.W. & Others, 2012).

FORMATION OF AN ADHESIVE BOND

A necessary condition for the creation of bonds is a proper approximation of atoms of the connected elements. In the case of adhesive bonding, adhesive is implemented between the two parts, which acts as an intermediate layer. The layer has an increased ability to create bonds with the atoms of the materials to be bonded (adhesive bonds), while maintaining sufficiently strong bonds between the atoms of this layer (cohesion force). During adhesive bonding, the adhesive layer undergoes physical or physico-chemical changes.

In order to obtain a durable connection, it is usually necessary to carefully **prepare the surfaces** of the connected elements, mainly by removing impurities and oxide layers. Sometimes a protection is also necessary when forming a connection against possible external influences. This applies in particular to joints resulting from thermal processes (Sobieszczański J., 2004).

Atmospheric environment has a very important but negative influence on the process of joining materials together. Oxygen is of particular importance since its presence in the atmosphere, oxide layers are formed on the surface of metallic materials. These oxides prevent the links between the atoms of combined metals from forming. Also, atoms of other gases can penetrate into the bonding zone, degrading the properties of the bond.

The basic mechanism of bonding is related to the bonding of an adhesive with a bonded material, which in turn takes place as a result of various adhesive processes. Mechanical adhesion is distinguished when the adhesive penetrates cavities or pores on the surface of the material. However, the adhesion associated with the combination of atoms and adhesive molecules with the material is essential. It can have a different character, because these linkages depend on the type of the adhesive and the material (Sobieszczański J., 2004).

The formation of adhesive bonds may facilitate the so-called **wettability**. Wettability is described as the ability of an adhesive to flow on a surface and the ability of atoms and adhesive molecules to penetrate into the structure of the base (bonded) material. Relations that result from the adhesion, form a permanent bond. However, in order for the bond to be practically applied, the bond between the adhesive molecules must also be preserved in the cured layer. These forces of internal cohesion are called cohesion forces. In practice, it is aimed to obtain the values of adhesion and cohesion forces similar to one another. It depends on their relation, in which area the connection will be destroyed, if the connection is loaded above the permissible value.

The strength of the bonded joint depends on the strength of the cohesion of the material and on adhesion forces occurring in the adhesive. Whether the glue will separate from the bonded elements or it will separate from the veneer itself or will the elements be destroyed depends on the ratio of the cohesion forces (Puff T. & Sołtys W., 1980).

An example of adhesive and cohesive joint failure is shown in Figure 6.

The problems of wettability, adhesion and cohesion are described in Chapter 3.

Figure 6. Joints damaged: a) adhesively (including adhesive, detached anodised layer),(b) cohesively

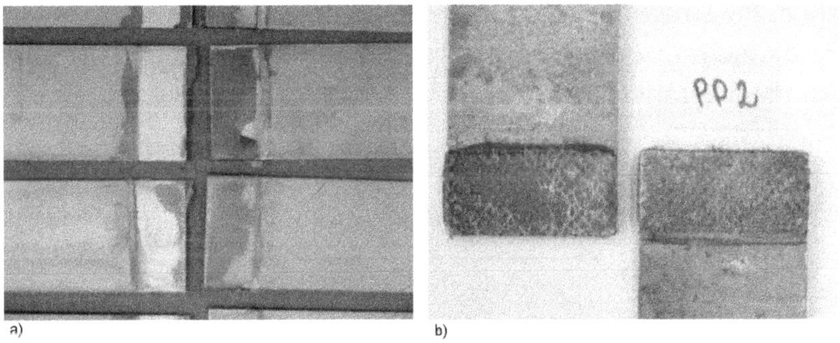

IDENTIFICATION OF ADHESIVE BONDS IN SPECIFICATIONS

In the technical drawing of the product specification, the adhesive bonding joint shall be marked in section, either by a thick line or from top view, by clearance. Also, a descriptive part shall be placed by the reference line finished with arrowhead. The descriptive part includes a graphic symbol of adhesive binding and the information about adhesive type. The reference line may indicate the location of the joint shown in a section or view perpendicular to the bonding plane (Figure 7).

Figure 7. Example of bonding specification (a), adhesive joint (b)

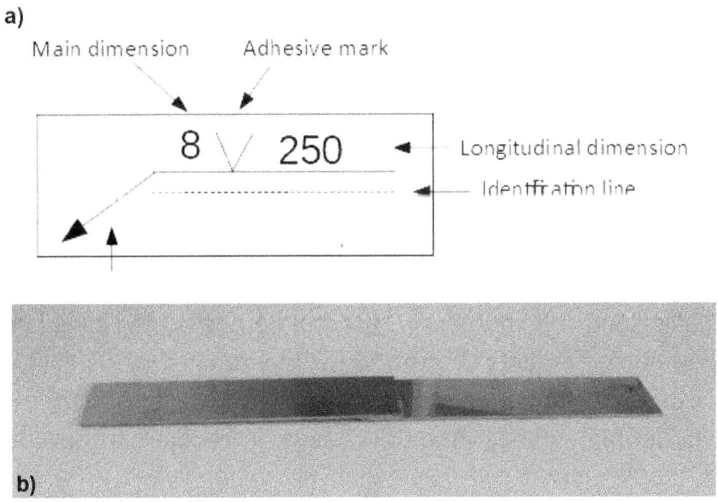

GUIDELINES FOR THE DESIGN OF ADHESIVE BONDS

In practice, the structures of adhesive bonding joints may vary. Most often they are divided into non-anchopic and tenon, the varieties of which are shown in the diagram (Figure 8).

Figure 8. Diagram - division of bonded joints

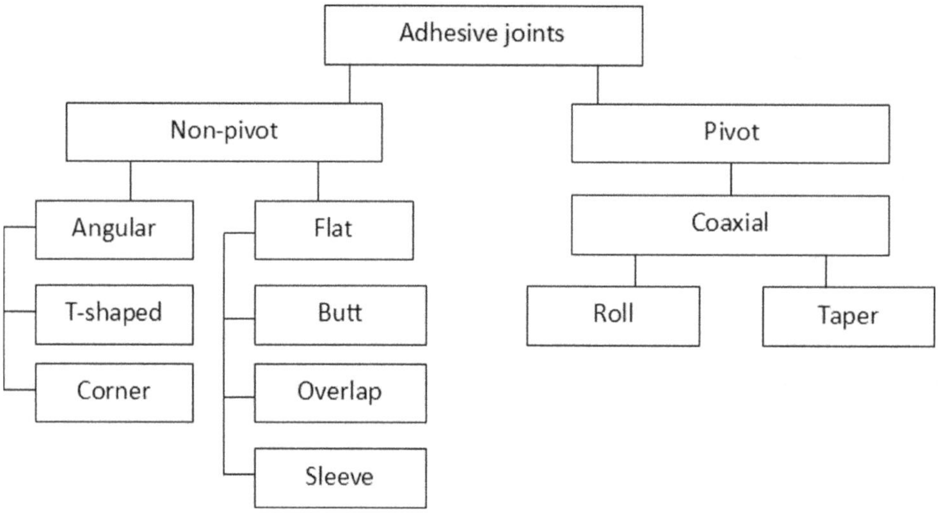

The strength of adhesive bonding joints is greater in shear than in tension and delamination, therefore the bonded joints should be constructed in such a way that they are exposed to shear and protected against delamination (Puff T. & Sołtys W., 1980). For this reason, overlapping or overlapping connections are most recommended. An additional normal force, the load on the joint creates compressive stresses in the joint and increases the ability to transfer shear stresses. The stress distribution in the overlap shear joint is uneven, however is smaller than in other types of permanent joints, e.g. welded joints. The highest stress values are found at the ends of the bond. These are the places of initiation of possible damage to the joint, therefore stress concentrations in these places should be reduced. This can be performed by a special way of shaping bonded elements (Figure 9) (Sobieszczański J., 2004).

Bevel cutting of bonded elements is a common construction technique used in overlapping joints, since this allows to reduce stresses, especially

Figure 9. Adhesive bonding joint structures: a) recommended, b) modified, correct, c) incorrect

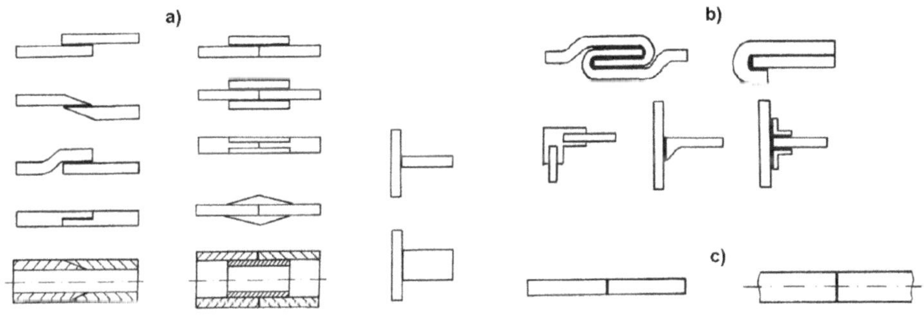

at the ends of the joint. As a result, load-bearing capacity of the joint is significantly increased.

The shear strength of a bonded joint depends significantly on the thickness g of the bonded elements and the ratio of the thickness g to the length of the l/g bond. At a constant value of the l/g ratio, increasing the thickness above a certain value results in a slight increase in strength, whereas at a decrease in thickness below this value, the strength is quickly reduced. If the edges are beveled, only a local additional increase in the deformation of the elements occurs, resulting in a reduction in the local deformation of the adhesive layer. If the thickness of the elements is reduced over the entire length of the joint, the increase in the deformation caused by the change in cross-section must cause an increase in shear stress on the adhesive surface and in the adhesive layer.

Sometimes, the size of the overlap affects the strength of the joint, although it is not correct to say that by increasing the surface area of the adhesive, the strength of the join will automatically increase. The opposite effect can sometimes be achieved, as confirmed by experimental studies described by (Kubit A. & Others, 2015), when the strength of single-overlapping joints deteriorated with the overlap growth.

CONCLUSION

The technique of adhesive bonding construction materials has a broad scope of application. From the overview carried out for this purpose, it is noted that there are currently various adhesives available on the market, the properties of which can be successfully selected to the materials to be bonded and the nature of the structure. In this context, it is essential to be able to select an appropriate adhesive, its additional components and to adapt them to the comprehensive technological expertise and production possibilities. Adhesive joints offer a number of advantages that distinguish bonding from other industrial bonding methods.

While on the one hand it could be assumed that the diversity works to the benefit of the manufacturers, on the other hand, however, a new analysis is required not only of the construction, the possibility of bonding it together, but also for the conditions in which it will be used.

REFERENCES

Brockmann W., Geiß, P.L., Klingen, J., & Schröder, B. (2009). *Adhesive bonding. Materials, applications and technology.* Weinheim: Wiley-VCH Verlag GmbH&Co. KGaA.

Cagle, Ch. V. (1977). *Kleje i klejenie* [Glues and Gluing]. Warszawa: Wydawnictwa Naukowo-Techniczne.

Miecińska, B., & Kubit, A. (2017). Analiza wybranych właściwości kompozycji klejowej z nanonapełniaczami [Analysis of selected properties of the adhesive composition with nanofillers]. *Technologia i Automatyzacja Montażu, 1*, 34–38.

Czaplicki, J., Ćwikliński, J., Godzimirski, J., & Konar, P. (1987): Klejenie tworzyw konstrukcyjnych [Bonding of construction materials]. Warszawa: Wydawnictwa Komunikacji i Łączności.

Ebnesajjad, S. (2008). *Adhesives Technology Handbook.* Norwich: Wiliam Andrew.

Farnell. (2019). Retrieved from https://pl.farnell.com/araldite/araldite-2021-1-50ml/epoxy-adhesive-cartridge-yellow/dp/2676048

Godzimirski, J., Komorek, A., Rośkowicz, M., Smal, T., & Tkaczuk, S. (2010). Tworzywa adhezyjne. Zastosowanie w naprawach sprzętu technicznego [Adhesive materials. Application in technical equipment repairs]. Warszawa: Wydawnictwa Naukowo-Techniczne.

Godzimirski, J., Kozakiewicz, J., Łunarski, J., & Zielecki, W. (1997). Konstrukcyjne połączenia klejowe elementów metalowych w budowie maszyn [Structural adhesive joints of metal elements in mechanical engineering]. Rzeszów: Oficyna Wydawnicza Politechniki Rzeszowskiej.

Habenicht, G. (2009). *Applied adhesive bonding. A practical Guide for Flawless Results*. Weinheim: Wiley-VCH Verlag GmbH&Co. KGaA.

Kelsall, R. W., Hamley, I. W., & Geoghegan, M. (2012). *Nanotechnologie*. Warszawa: PWN.

Kniewald, D., & Baron, A. (1994). Pretreatment of the substrate surface prior to bond. *Technologia i Automatyzacja Montażu*, *2*, 9–12.

Kubit, A., Ciecińska, B., & Drozd, K. (2015). The influence of lap size on shear strength of adhesive joints. *Advances in Sciences and Technology*, *9*(26).

Marcq, F., Demont, P., Monfraix, P., Peigney, A., Laurent, C., Falat, T., ... Jamin, T. (2011). Carbon nanotubes and silver flakes filled epoxy resin for new hybrid conductive adhesives. *Microelectronics and Reliability*, *51*(7), 1230–1234. doi:10.1016/j.microrel.2011.03.020

Porębska, M., & Skorupa, A. (1997). Połączenia spójnościowe [Coherent connections]. Warszawa: Wydawnictwa Naukowe PWN.

Puff, T., & Sołtys, W. (1980). Podstawy technologii montażu maszyn i urządzeń [Fundamentals of machine and device assembly technology]. Warszawa: Wydawnictwa Naukowo-Techniczne.

Rośkowicz, M. (2013) Wpływ lepkosprężystości klejów konstrukcyjnych na wytrzymałość połączeń klejowych [Impact of viscoelasticity of construction adhesives on the strength of adhesive joints]. Warszawa: Wojskowa Akademia Techniczna.

Sobieszczański, J. (2004). *Spajanie [Joining]*. Warszawa: Oficyna Wydawnicza Politechniki Warszawskiej.

Zielecki, W., & Kubit, A. (2013). Wpływ proszkowych nanonapełniaczy ceramicznych na wytrzymałość statyczną połączeń klejowych [The effect of ceramic powder nanofillers on the static strength of adhesive joints]. *Technologia i Automatyzacja Montażu*, *4*, 45–48.

Chapter 2
Construction Materials for Adhesive Bonding in Present-Day Industry

ABSTRACT

This chapter presents the characteristics of selected construction materials used in the modern industry, with particular attention paid to the engineering industry. These materials have a diverse chemical structure and related properties. In this context, steels, selected light alloys, plastics, wood, glass, and ceramics were described. Since the materials for production are supplied in various forms, the methods of shaping the semi-finished products of the presented materials were outlined. Special attention was given to the condition of the surface and the shape due to the possible bonding operations in the technological process. Additionally, a description of composites was given, as these materials have already replaced a number of materials used so far. Composites are characterized by having favorable features and are produced with the use epoxy resins, among others. This chapter also outlines the characteristics of adhesives in the context of their selection options for the glued materials.

DOI: 10.4018/978-1-7998-4634-5.ch002

INTRODUCTION

Currently, numerous diverse structures are applied in a variety of industries. They are designed and made from various materials of chemical structure and properties customized to match the demands. Several main material groups used in industry can be distinguished in order to manufacture various products. These products can be then successfully bonded as a result of modern substances used for adhesive bonding. A distinction of a basic type of materials should be drawn in this group, which includes, for example: steel, light alloys, including: aluminum, titanium, beryllium and magnesium, plastics, glass and ceramics, wood, composites, etc. (Dobrzański L.A., 2002; Virgamet, 2019).

STEELS

Steels are iron alloys with carbon content not exceeding 2%. According to their chemical composition and varied properties they can be divided into two basic groups (Buchnik G., 2006):

- **carbon steels**, with admixtures of manganese (max 0,8%), silicon (max 0,4%), phosphorus (max 0,05%) and sulfur (max 0,05%). Additionally, carbon steels may contain small quantities of copper, chromium and nickel resulting from steel scrap, as well as dissolved gases, such as hydrogen, oxygen and nitrogen. In carbon steels, increased contents of carbon increase tensile strength Rm, yield point Re, and hardness; decreases plastic properties – elongation A, contraction Z and impact strength KC. However, the increased carbon content impairs weldability and machinability. In turn, manganese, which appears to be a smelting process residue, enhances steel durability and positively affects weldability. Silicon found in carbon steels comes from deoxidation processes. It enhances steel durability, particularly the elasticity limit, yet deteriorates its weldability. Another harmful admixture is phosphorus. Although it increases tensile strength and hardness, it considerably decreases plastic properties. It also raises the temperature at which steel becomes brittle. In rare instances phosphorus is a desirable admixture when it is supposed to enhance machinability, and with accompanying copper presence it improves corrosion resistance. Similarly to phosphorus, sulfur is regarded as a harmful admixture. It does not dissolve in iron and remains in the form of iron and manganese sulfides. During hot machining, iron sulfides (FeS) cause cracks and tears (at above 800°C). Manganese sulfide (MnS) is hardly fusible and occurs in the form of separate inclusions. With sufficient quantity of manganese, iron sulfides (FeS) are not likely to occur. Among other admixtures, oxygen and hydrogen cause steel brittleness in hot machining (like sulfur), while nitrogen manifests itself as hard and brittle nitrides. Types of carbon steels are divided into three categories: construction steels of different grades, tool steels and those of special properties (easily machinable, free-cutting steels) (Table 1) (Dobrzański L.A., 2002; Rudnik S., 1996).

Table 1. Carbon steels and their uses

Steel Grade	Symbol	US Standard*	Uses
Low carbon steel D11	A 366(C)	ASTM A 366-97	Various structural elements
Machining steel for hardening and tempering C45	~1045	ASTM A 29-99	Axles, shafts, spindles, rollers, pump impellers
Free-cutting steel 11SMn30	~1213	SAE J 403-1995	Parts manufactured in high speed machining
Machining steel for hardening and tempering C60E	~1060	ASTM A 29-99	Heavily loaded and abrasion-resistant machine parts, spindles, rollers, rings, axles
Spring steel 65	~1070	ASTM A 713-93	Springs, shock absorbers, line spring elements
Tool steel N11	W1-10˜C	ASTM A 686-92	Forming dies, punch tools, cutting tools, tool chuck elements

* ASTM – American Society for Testing and Materials
SAE – Society of Automotive Engineers

- **alloy steels**: include structural steels (low-alloy steels, spring steels, steels for carburizing, steels for hardening and tempering, steels for nitriding, rolling bearing steels), tool steels (for cold and hot working, high-speed steels), as well as those characterized by special properties (for working at higher temperatures, for pipes, metal sheets, bolts, forgings), corrosion resistant, heat resistant, with special physical and magnetic properties.

Structural alloy steels are used to manufacture various elements working at temperatures between -40 and 300°C in environments which do not affect them adversely. Alloy additives, among other things, enhance tempering and grain refinement. Such steel wares can withstand static and dynamic loads. Tool steels are intended to be made into machining and plastic working tools, as well as measuring instruments. They are characterized by high hardness, abrasion resistance, tempering, and resistance to high temperatures.

Stainless and acid-resistant steels are materials whose chemical resistance depends on alloying additions, i.e. chromium, nickel, carbon, molybdenum, copper, manganese, nitrogen, titanium, niobium and tantalum, where chromium is a primary addition.

Heat resistant steels, with chromium, silicon and aluminum additions, are intended to be made into furnace fittings, electric power equipment, jet engine parts due to their resistance to oxidizing effects of gases at temperatures above 550°C.

Steels with high electrical resistance used as resistance heating elements are mainly nickel-chromium alloys, chrome-nickel steels and chrome-aluminum steels.

Steels and their alloys, characterized by special magnetic properties, are used to produce power generator and transformer plates, magnets – it could be pure iron, but also low carbon steels, silicon steels and nickel iron alloys.

In turn, chromium-nickel-manganese steels (H12N11G6) and chromium-manganese steels (G18H3) (Table 2) (Przybyłowicz K., 2012; Blicharski M., 2004) are considered to be non-magnetic materials or indifferent to magnetic field.

LIGHT ALLOYS

Metals, such as titanium, aluminum, magnesium and their alloys can successfully replace steel due to their properties. The development of manufacturing techniques and better understanding of material properties enables the use of light alloys in industries where not only the low weight of element but also its high reliability, resistance to various factors (high or low temperatures, chemical substances) and rigidity play an important role. Light

Table 2. Alloy steels and their uses

Steel Grade	Symbol	US Standard*	Uses
Machining steel for hardening and tempering 25CrMo4	4130	ASTM A 322-91	Heavy machinery parts, electric Power equipment, aircraft structural components, drive shafts
Spring steel 50HF	~6150H	ASTM A 304-95	Leaf springs, shock absorbers, choke springs, cable connectors
Structural steel for carburizing 16HG	~5120H	ASTM A 534-94	Transmission gears, camshafts, dowel pins, helix gears
Corrosion resistant steel X6Cr17	~430	ASTM A 240-56	Nitric acid production equipment, tanks, sanitary fittings and tubing, pipelines
Heat resistant steel H16N36S2	~330	ASTM A 554-94	Parts for heat treatment furnaces, protective atmosphere generators
Refractory steel H5M	~501	ASTM A 193-96	Crude oil processing apparatus, pump components, boilers, boiler covers, turbine blades
Valve steel H9S2	~HNV3	SAE J 775	Engine exhaust valves for cars and motorbikes
High speed steel SW7M	M2	ASTM A 600-92	High-strength machining tools, grill bits, saw blades, threading taps and dies, broaches, lathe cutting tools
Bearing steel ŁH15	52100	ASTM A 535-85	Rings, rollers, steel balls, roller bearing needles
* ASTM – American Society for Testing and Materials SAE – Society of Automotive Engineers			

alloys are currently used in aircraft manufacturing, aerospace and automotive industries, food processing industry, pharmaceutical industry and others.

Titanium and its Alloys

Titanium, as a construction material, is preferably used given its high mechanical strength and low density 4,4 of g/cm^3. It is lighter than steel by approx. 45% but 60% heavier than aluminum alloys. In addition, it has very good chemical properties, including high resistance to atmospheric corrosion, sea water and organic acid corrosion (however, it is not resistant to HF, H_2SO_4 or HNO_3), low thermal and electrical conductivity. It is used in aviation and aerospace industries to make wing support structures, connecting elements (rivets, bolts) and actuators (Bielawski R. & Others, 2015). In titanium alloys, the main alloying elements are aluminum (Al), tin (Sn), molybdenum (Mo), vanadium (W), manganese (Mn), iron (Fe), chromium (Cr). The alloying additions can improve properties of the material (e.g. mechanical properties), when the titanium alloy reaches a specific strength equal to that of improved alloy steels, or even higher. Alloys with such characteristics are used in the construction of spacecraft structures, components of compressors and turbines. Titanium alloys are used in the production of chemical apparatus components, surgical instruments and medical implants due to their excellent chemical resistance (Table 3). In addition, some of them show the so-called shape memory effect (Przybyłowicz K., 2012; Dydek Ł. & Others, 2016; Przybyłowicz K., 2012).

Table 3. The selected titanium alloys and their uses

Titanium alloys	Symbol	US Standard*	Uses
Titanium Ti99	R50700	ASTM B 265-1995	Construction elements in aviation, aerospace, chemical and optical industries, medical implants and prosthetics
Titanium alloy Ti6Al4V	R56400	ASTM B 265-1995	Rocket engine shells, turbine engine components, rings, blades, aircraft hydraulic fittings parts, pressure vessels
* ASTM – American Society for Testing and Materials SAE – Society of Automotive Engineers			

Aluminum and its Alloys

Aluminum alloys are construction materials extremely widespread in many industries. Pure aluminum has a low density of 2.7 g/cm^3, and after supersaturation and ageing, when precipitation hardening (also called dispersion hardening) occurs, it obtains high mechanical strength of 400 MPa approx. Aluminum is resistant to water, H_2CO_3, H_2S, numerous organic acids and nitrogen compounds, it is not yet resistant to hydroxides (e.g. NaOH, KOH), anaerobic acids (HF, HCl), sea water or mercury ions (Przybyłowicz K., 2012).

The main additions in aluminum alloys are copper, silicon, magnesium, manganese and zinc. Small amounts of nickel, iron, titanium, chromium and beryllium are also added, producing varieties of alloys with specific properties. Aluminum alloys are used in two variants – alloys for plastic forming, resulting in intermediate products of parts suitable for gluing, and less frequently casting alloys bonded by means of adhesives (Kuczmaszewski J., 2011).

Aluminum alloys are used for heavily loaded elements of aircraft structures, e.g. spars, ribs, skin and construction brackets. Often, such elements are joined with rivets or rivet nuts also made of aluminum alloys (Bielawski R. & Others, 2015; Kuczmaszewski J., 2011). In addition, aluminum alloys are used in manufacturing of compressor blades, fan discs adapted to work at temperatures from 200 to 300°C, fuel tanks and oil lines (Oczoś K., Kawalec A., 2012; Aluminum, 2020).

Selected aluminum alloys are shown in Table 4.

Table 4. Selected aluminum alloys and their uses

Aluminum alloy	Symbol	US Standard*	Uses
Aluminum	1050A	AA Reg. Rec.	Containers for food, chemical and brewing industries
Al-Cu alloy**	2017A		Corrosion resistant elements, automobile components, aircraft constructions
Al-Mn alloy**	3004		Containers and cans, shipbuilding and chemical industry components
Al-Si alloy**	4032		Combustion engine pistons, components of devices operating at elevated temperatures, corrosion resistant components
Al-Mg alloy**	5052		Components for food, chemical, aircraft and construction industries, sea transport, oil installations, fuel tanks
Al.-Mg-Si alloy**	6063		Decorative elements in construction industry, fittings, motor vehicle and railroad car flooring
Al.-Zn alloy**	7049A		Heavily loaded elements of aircraft and military constructions
Other alloys, e.g. Al-Li**	8090		Aircraft and aerospace construction elements exposed to cryogenic conditions
* AA – Aluminum Association in USA ** The alloy also contains other elements			

Magnesium and its Alloys

Magnesium is one of the lightest metals and has a density of 1.74 g/cm^3 (approx. 5 times lower than iron). Pure magnesium is rarely used in industry due to its susceptibility to oxidation, low hardness, low strength and plasticity. One of the greatest problems with magnesium is its poor corrosion resistance. Therefore, it is necessary to apply a protective coating, e.g. powder coating, nickel and chrome galvanizing or anodizing.

A variety of magnesium alloys is used in the industrial field mainly with aluminum, zinc, manganese, zirconium or silicon. In recent years, ultra-light magnesium-lithium alloys have also appeared (Oczoś K., Kawalec A., 2012).

Similar to aluminum alloys, magnesium alloys are made in two varieties: alloys for plastic working and foundry alloys. In order to increase the mechanical properties of elements operating at elevated temperatures, cerium, thorium, lanthanum and neodymium are added. Other alloying elements might include aluminum, zinc, manganese, lithium, beryllium, silver, tin and zirconium (Oczoś K., Kawalec A., 2012; Dobrzański L. A., 2001; Westegen H., 1993; Kuczmaszewski J., 2011).

Because of their low weight and other properties (magnesium alloys are alkali and temperature resistant, within operating temperature range of -190 to +250°C), they are used in the aircraft industry for heavily loaded structural elements, such as airframes, tanks, fuel and oil lines, covers. Magnesium alloys are also used to produce other, less loaded structural elements, such as space frames, gear housings and covers, wheel components. They are also used to make electronic device housings (notebooks, cameras, cell phones) (Table 5) (Dobrzański L. A., 2001; Oczoś K., Kawalec A., 2012).

Table 5. Selected magnesium alloys and their uses

Type	Symbol	US Standard*	Uses
Pure magnesium Mg	9990 B	ASTM B 92-1994	Alloys
Magnesium alloy	AZ 63	ASTM B 403-1996	Parts of airplanes, cars and engines, instrument and camera housings
Magnesium alloy	AM100A	ASTM B 80-1993	Photo and video camera components, tools
Magnesium alloy	ZK 51 A	ASTM B 80-93	Elements of aircraft structures, wing skins, helicopter components

* ASTM – American Society for Testing and Materials

Beryllium and Beryllium Alloys

Beryllium has the lowest density among light metals (1,848 g/cm^3) and has different physical properties unlike all other metals. It is characterized by a relatively high melting point, large modulus of elasticity, high electrical conductivity, high specific heat as well as low coefficient of thermal expansion. Beryllium is 1.5 times harder than steel, 2.5 times harder than titanium and

4 times harder than aluminum and at the same time is a brittle and non-magnetic material.

Wrought beryllium has the highest specific strength among structural metals and retains it to the temperature up to 600°C approximately. Mechanical properties of beryllium depend on the method of manufacturing and further machining and thermal treatment. In addition, beryllium has anisotropic properties after rolling and extrusion.

In dry air up to 600°C beryllium does not oxidize, however rapid oxidation occurs at between 900 and 1000°C. At temperatures above 600°C it reacts slowly with nitrogen and forms beryllium nitride (Be_3N_2).

At temperatures up to 100°C it is stable in pure water, whereas when put in water with chlorides or metal carbides, it corrodes quickly.

Melted beryllium reacts with carbon and with persistent and flame retardant oxides such as magnesium oxide (MgO), aluminum oxide (Al_2O_3), calcium oxide (CaO) and zinc peroxide (ZnO_2).

From beryllium and its alloys rods and pipes are produced, however it is a toxic material, expensive in manufacturing and is of limited use.

Beryllium alloys are used in various fields (Oczoś K., Kawalec A., 2012; Haws W. J., 2000); Materion, 2019):

- with Be-Al aluminum used in optics and telecommunication, in structures with adhesive bonds exposed to high temperatures, where sensitive aluminum alloys are replaced, in aviation and aerospace instead of aluminum and titanium alloys,
- with Cu-Be copper as high-strength alloys (with Be content of 1.6 to 2%) and alloys with high conductivity (approx. 3% Be content), which are used in electronics, automotive, household products, precision instruments, thermostats, jet engines, telecommunications due to their good corrosion resistance, non-magnetic properties and high fatigue strength,
- with Ni-Be nickel with good elastic properties at elevated temperature, used in diaphragms, thermostats, connection sockets,
- other, e.g. composites of beryllium and beryllium oxide used in electronics.

PLASTICS

There is a substantial group of plastics among construction materials. Plastic is the name usually given to a material whose basic component is a polymer, or macromolecular compound obtained by chemical treatments. Sometimes plastic consists of only one polymer. Most often, however, other substances used for various purposes are added to the polymer. Polymers can be divided into two groups: polymers which occur in a natural form, e.g. rubber, cellulose and starch and polymers which are synthetically obtained by chemical synthesis of compounds called monomers. Synthetic polymers can be organic or inorganic compounds. Polymers may also include elements such as oxygen (O), chlorine (Cl), fluorine (F), silicon (Si) and sulfur (S). Molecular weight of the polymer determines a number of its physicochemical and mechanical properties such as melting point and plasticity temperature, solubility, tensile strength, resistance to multiple deformations, modulus of elasticity, chemical and thermal resistance, and others (Żuchowska D., 2000).

To polymers are added some chemicals to modify their physical or chemical properties. These can be extenders, carriers, stabilizers, plasticizers (softeners), antioxidants, dyes, pigments, destaticizers, etc. (Żuchowska D., 2000).

The properties of plastics can be modified by chemical or physical techniques. The purpose of polymer reinforcement is to improve its rigidity and dimensional stability, ability to suppress mechanical vibrations, heat and chemical resistance, increase gas or liquid permeability, improve their processing properties or economic indicators. Often, such modifications result in the improved expected property, yet can also deteriorate another, thus a compromise solution is most often chosen in a production process (Żuchowska D., 2000).

There are many ways of classifying plastics, depending on the criterion adopted. The most common classification which is applied is rheological, as they are related to the characteristics of performance. According to such criteria a distinction is made between elastomers which show a large deformation at low stresses, even up to 1000%, and plastomers which show a very low deformation (less than 1%) under stress. Plastomers are divided into thermoplastics and duroplasts, which in turn can be thermosetting or chemically cured. However, this division is not rigid due to the fact that many mixed plastics are produced to obtain specific properties, e.g. elastomer reinforced thermoplastic polymers (Dobrzański L. A.).

One of the undisputed advantages of plastics is their low density, high mechanical strength in relation to weight, durability, weather resistance, durability in chemically active environments, dimensional stability at elevated temperatures, specific electrical or thermal properties (conductivity or lack thereof), easy dyeing in different colors or the use of transparent or translucent materials.

From the point of view of manufacturing processes, the advantageous feature is the possibility of shaping finished products in relatively simple technological processes.

When choosing plastics as construction materials, however, it is necessary to take into account features that might prove unfavorable, e.g. low creep resistance, limited temperature range for long-term use, water swelling or brittleness at low temperatures. Plastics, however, are a group of materials that is under continuous development. This is not only due to possible additions and modifications of their performance properties, but also because of environmental problems and the need for reprocessing and utilization of recycled products. It is a serious problem due to a huge variety of substances added to polymers, their structure (powders, granules, fibers, lamellae, porous or smooth texture) and properties.

The examples of most common plastics used in machine constructions are presented in Table 6-8 (Dobrzański L. A., 2008; Plast, 2019).

Table 6. Selected most common thermoplastics and their uses

<table>
<tr><th colspan="2">Thermoplastics</th></tr>
<tr><td>Basic properties</td><td>Rigid at room temperature
Softens reversibly when heated</td></tr>
<tr><td colspan="2">Primary examples and application
ABS – acrylonitrile-butadiene-styrene – housings for household appliances and agricultural equipment, automotive components, toys
PA – polyamide – bearings, gears, handles and cable covers
PC – polycarbonate – protective helmets, camera lenses
PE – polyethylene – bottles, containers, toys, components of kitchen utensils
PET – polyester –beverage bottles, containers, housings, fabrics
PMMA – polymethyl methacrylate – transparent elements of aircraft components, road signs, lenses
PP – polypropylene – bottles, food packaging boxes, device casings
PPS - polyphenylene sulfide – household appliance housings, parts for automotive industry, transporting equipment components, electronics, packaging boxes, laboratory and medical equipment components
PS – polystyrene – battery casings, packaging boxes, toys
PTFE – polytetrafluoroethylene - bearings, pipes, non-stick coatings, electronic parts, elements of aircraft components
PVC – polyvinyl chloride – components requiring solvent resistance in various industries
SAN – poly(styrene-co-acrylonitrile) – casings, containers of various types, casings, containers of various types, finishing and decorative elements</td></tr>
</table>

Table 7. Selected most common elastomers and their uses

Elastomers	
Basic properties	Flexible at room temperature Shows increased tendency to flow when heated
Primary examples and application **Natural rubber:** NR – natural rubber – tires, hoses, shoe soles, seals and washers **Synthetic rubber:** CR – chloroprene rubber – wire insulation, hoses, seals and washers, chemical tank liners NBR – acrylonitrile butadiene rubber – fuel, oil and petroleum hoses, seals and washers VMQ – silicone rubber – thermal insulation materials, membranes, seals, medical devices, food packaging	

Table 8. Selected most common duroplasts and their uses

Duroplasts	
Basic properties	Rigid at room temperature Once molded retain their shape despite heating
Primary examples and application EP – epoxy resin – electronic components, protective coatings, adhesives, laminates PF – phenol-formaldehyde resin – housings for cell phones, motors and other electrical equipment UF – urea-formaldehyde resin – household appliances, plugs and other connectors for electronics and electrical engineering UP – unsaturated polyester – automotive and aircraft components, interior design materials, fan propellers, fiberglass boats, safety helmets, sports equipment MF – melamine-formaldehyde resin - decorative laminates, decorative laminates, electrical accessories (sockets, outlets, switches)	

Plastics can be bonded, yet the process should be adapted individually to the type of plastic and the selection of an appropriate adhesive should be taken into account. The most widely used is PVC and PS gluing (Krywult B., 1990).

WOOD

Wood is the oldest construction material and is still used in many structures. It is lightweight, sturdy, as well as aesthetic and durable. On account of these specific characteristics, it is used not only in a broad range of industries, but also in the production of decorative elements, everyday objects and art supplies. Sometimes wood is regarded as a natural and versatile fiber reinforced composite material (Dobrzański, 2008) with anisotropic mechanical properties, depending on the direction of the fibers (it is most durable in longitudinal direction). It is a material with longitudinally arranged tubular polymer cells in a polymer matrix, whose pores are filled with air and water with a variable composition. Wood as a construction material is obtained from various tree species, whose timber may contain resin, tannins and essential oils. These components affect the functional properties of wood (hardness, machinability, hygroscopicity and color). Wood from deciduous and coniferous trees is called hardwood and softwood respectively; however, it does not literally refer to its hardness. Timber, with respect to its seasoning stage, can be classified as wet or dried (Table 9).

In the raw state, without any additional protection in the form of waterproofing chemicals or coatings, wood has a low density, durability and elasticity. It also has a low coefficient of thermal expansion. However, for manufacturing needs it has sufficient strength, low thermal conductivity and is easy to machine.

When choosing wood for production one should however remember about the risk of punkwood, damage of insect habitats, e.g. bark beetles, or plant tissue damage (Table 9) (Dobrzański L. A., 2008; Las, 2020).

GLASS AND CERAMICS

Glass is defined as an inorganic, amorphous solid which does not have ordered internal structure or has no constant melting point and is a poor conductor of electricity, while it is chemically resistant, except for hydrofluoric acid. Properties of glass depend on the method of melting and its substrate composition.

The composition of glass, in addition to glass-forming components such as quartz sand (SiO_2) and additives in the form of boron oxide (B_2O_3), germanium oxide and (GeO_2) may include: sodium carbonate (Na_2CO_3), calcium carbonate ($CaCO_3$), lead oxide (PbO) flux and pigments. The glass color depends on the type of metal oxide, thus green glass contains iron (II) and chromium (III) compounds, blue glass contains cobalt compounds, purple glass manganese compounds, yellow glass cadmium and sulfur compounds, red glass colloidal gold particles.

The ingredients are mixed, melted in an oven at 1200 to 1300°C and formed before solidification.

There are several types of glass, according to their composition and use: quartz glass (with quartz as the main ingredient), which is used in the production of laboratory equipment due to its high resistance to sudden temperature changes, acids, as well as UV transmittance; leaded glass (with

Table 9. Most common timber species and their uses

Timber	Wood color	Uses
Softwood		
Alder	Orange-red	Plywood, pencils, drawing supplies, foundry models
Balsa	White	Structural components of floating vessels (rafts, yachts), airplanes and gliders, deck equipment, water rescue equipment, surfboards, orthopedic prosthetic devices, model airplane kits, conductor batons (due to extremely low density)
Birch wood	White with a slight yellow tinge	Plywood (for aircraft industry, carpentry and furniture making), household goods and decorative products
Linden	Whitish yellow, glossy	Foundry models, drawing boards, model airplane and ship kits, sculpting material
Pear tree wood	Light to dark orange	Firearm stocks, foundry models, drawing supplies, sculpting material
Pine wood	Pale yellow to reddish brown	Building poles, railroad ties, plywood, woodworking material, paper production
Poplar wood	White or pale gray with a greenish or yellowish grey tinge	Matches, barrels, boxes, toys, pencils, home accessories, components of musical instruments (guitars and drums)
Hardwood		
Acacia/robinia	White with a slightly yellow tinge	Tool handles (hammers, axes, pickaxes)
Ash wood	Light yellow	Floor staves, wall panels, furniture, skis and aircraft propellers
Beech wood	Pale cream with a pink tinge	Aircraft plywood, floor staves, furniture, barrels and small construction elements
Elm wood	Light to medium dark brown	Boatbuilding components, furniture
Hornbeam wood	White to pale yellowish brown	Components of agricultural machinery, sleigh runners, boat components, blocks, piano action parts
Oak wood	Nearly white to a darker reddish brown	Furniture and home accessories, flooring, wine barrels, window frames
Sycamore wood	White to pale colouring	Violins, floor staves, small decorative items, sculpting material

lead oxide as an additive), used for the production of optical lenses, and protective shielding against x-ray and gamma radiation; soda-lime glass (with CaO, SiO_2, Na_2O as additives) used for window panes, glass containers and light bulbs; architectural glass (soda-lime-silica glass) used for window panes, electric stove tops, fireplace doors, decorative elements in various colors, plain or reinforced, electrically conductive, transparent or opaque. In recent years, toughened glass panes have also been glued to aluminum housings in solar panels (Bernaciak M., 2010).

The architectural glass group also includes glass blocks and glass wool used for insulation purposes. In recent years, new forms of glass have been produced. These are glass fibers, formed by squeezing liquid glass through the eyelets of a molding machine, with a diameter of 3 to 20 μm. These fibers, of various compositions, are characterized by favorable mechanical properties and are used to manufacture composite materials by epoxy resin gluing. These materials, also called laminates, have been used, among others in the aircraft, defense and chemical industries (Dobrzański L. A., 2008).

It is worth to mention glass ceramics formed by a controlled crystallization of glass. Glass ceramics are characterized by high abrasion and thermal shock resistance and mechanical properties higher than those of glass. All these characteristics make glass ceramics widely used in metallurgical furnace constructions, baking pans and molds, abrasion resistant flooring, weaponry items, telescopic mirrors and high-performance tools.

SEMI-FINISHED CONSTRUCTION MATERIALS

The materials presented above are used in production processes as semi-finished products. There are several basic groups of semi-finished products: castings, forgings, metallurgical materials, die-stamping products.

However, from the point of view of gluing operations, shaped materials are useful to obtain flat or rounded, regular surfaces that can be prepared by mechanical or chemical conventional methods, which are explained further in this work. The form of a semi-finished product should be selected so as to obtain a product as close to the finished product as possible, ensuring proper machining and operational properties expected from a finished product at the same time. These conditions should be met by the material grade, its properties, shape and dimensional accuracy as well as surface roughness.

A semi-finished product may have surfaces intended for further shaping during technological processes. These surfaces should have the quality required for gluing operations (Łunarski J., Banaś G., 1988).

Metallurgical Semi-Finished Products

Metallurgical semi-finished products include rolled, drawn, peeled and ground products of various shapes – flat (bars, sheets), structural sections: angles, U-, T- and I-sections; square, rectangular, hexagonal and round bars; pipes.

Rolled rods, which are obtained by cold or hot forming, are made of various metals, such as low-alloy carbon structural steels, high speed steels and high alloy steels. The type of the material affects the size of the rods produced. They are usually provided by manufacturers and made available according to country-specific standards. The dimensional accuracy of rolled bars is relatively low (IT15-IT16) (Wodecki J., 2011).

Bars can be produced in a more accurate way, by means of pulling. In this way, their dimensional accuracy stays within IT9-IT13. Such rods can be additionally ground or polished. A variation of a surface finish finishing process of bars is peeling, yet it provides a lower accuracy of IT11-IT16. Sometimes, in order to improve properties and performance, such as fatigue or abrasion resistance, bars are roller burnished. Drawn and additionally machined bars have a relatively high dimensional and form accuracy as well as a particularly good surface quality. Such rods often require no additional processing, except for special finishing processes, like coatings (protective, aesthetic appeal), or assembly, e.g. by gluing. Then, gluing procedure is followed, removing possible contamination at first to improve adhesion conditions.

Similar to rods, sheets and pipes are most often hot rolled.

Sheet metal is delivered in various sizes and thicknesses. Sometimes, dimensions are standardized, recommended by the manufacturer. Structural elements made of sheet metal can be combined by welding, gluing, soldering, screwing or riveting, and finished by sealing, painting, marking, etc. Sheet metal can also be used to manufacture products formed by extrusion.

Pipes are produced in seamless and seamed versions. The seamed pipes are made by welding metal strips or sheets.

In the field of adhesive bonding, wires constitute a less useful form.

In the case of metallurgical materials, their surface quality is an important factor in adhesive bonding. Due to the fact that the material is often hot-formed, the surface of the semi-finished products may have pitting corrosion centers. Sometimes cases of partial melting may also occur. Such defects should be removed before bonding, e.g. by cutting or shot blasting (Figure 1).

The examples of types of materials are shown in Table 10 and Figure 2.

Patterns, Pressings and Stampings

The metallurgical materials described above, such as flat bars, sheets, pipes, are made into punched patterns. Machining of the starting metallurgical material involves cutting out specific shapes of elements which may be combined

Figure 1. The example of a corroded part (a) cleaned by shot blasting (b)

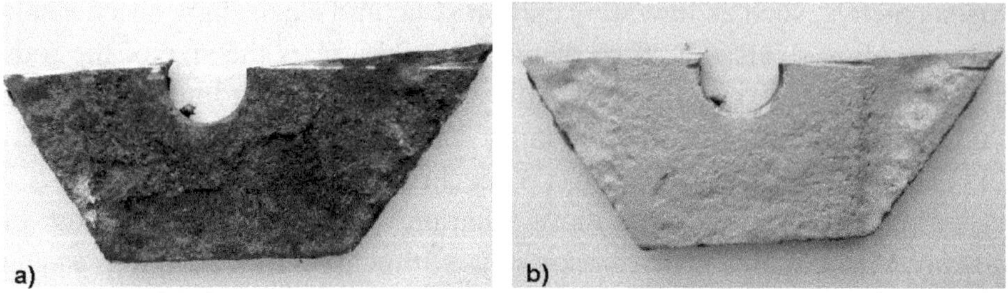

Table 10. List different materials forming

Material	Semi-finished product
Steels	Sheets, bars of various cross-sections, wires, pipes, forgings, castings
Titanium and its alloys	Pipes, sheets, wires, rods
Aluminum and its alloys	Sheets, foils, rods, profiles, pipes, wire, castings
Magnesium and its alloys	Sheets, pipes, rods and sections (forgings)
Plastics	Moldings: rods, plates, pipes, foils, fibers; castings
Wood	Sawn timber (work boards: profiles, flat bars, angles), circular sections made using cutting machines (as opposed to rolling)

by multiple techniques. This method of preparing semi-finished products is often used in the production of a small range of products (prototypes, unitary production) due to a relatively high labor intensity when cutting with guillotines or shears. It is also applied in mass production where drilling machines, multi-knife heads, acetylene and jet plasma burners are used, and the material is cut with a laser or a water jet. Smaller elements are also cut using presses with the use of punching dies. In the cutting methods described above, both efficiency of the cutting process and quality of the obtained edge is high (Feld M., 2009).

Metallurgical materials, from which semi-finished products were made earlier, may be used to make semi-finished products or finished products by cold forming. Pressing, die stamping, spin forming, drawing are shape changing methods of, e.g. previously prepared pattern. Such treatment is used in justified cases in order to save the time necessary to produce the goods (combined plastic forming and machining is possible), and to offer material savings. Extrusions can be bonded in various techniques at subsequent stages of the technological process by means of gluing, painting, varnishing, etc.

Figure 2. Semi-finished products of metallurgical materials: a) square tube, C-profile, I-profile and angle section - steel, b) round and square tube, round tube and hexagonal tube - steel, c) tube, wire and angle section – aluminum

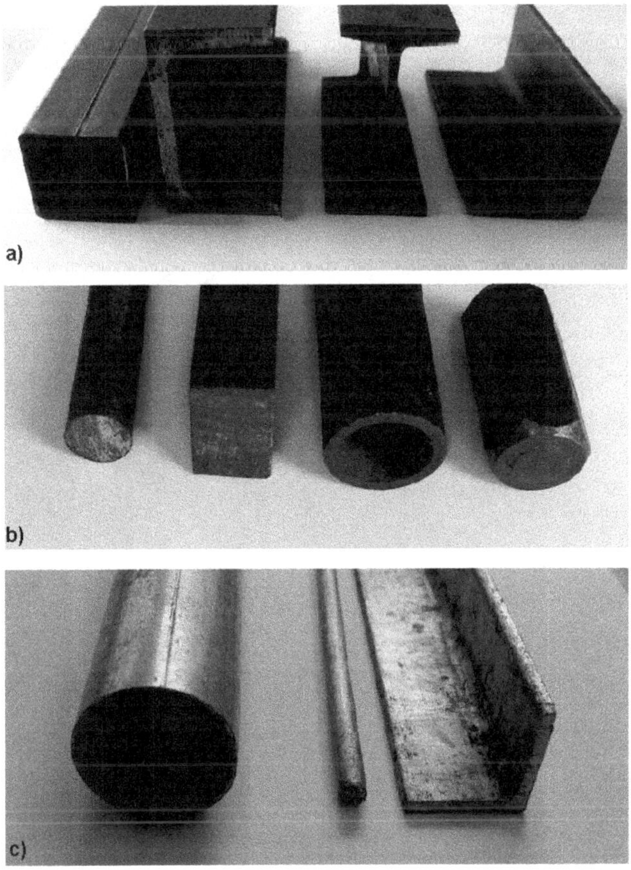

The groups of pressings may also include plastic moldings which are pressed from granules or properly prepared powders (selected in terms of composition and mixed). In this way, various types of fittings are obtained, corresponding to the shape of the cross section of the die. Sometimes plastic moldings resemble metal parts made by rolling. As in the group of metallurgical materials, bars of different cross-sections, plates, tapes and tubes can be manufactured (Krywult B., 1990).

Figure 3. Sheet metal cuttings and moldings

Figure 4. Plastic extrudates

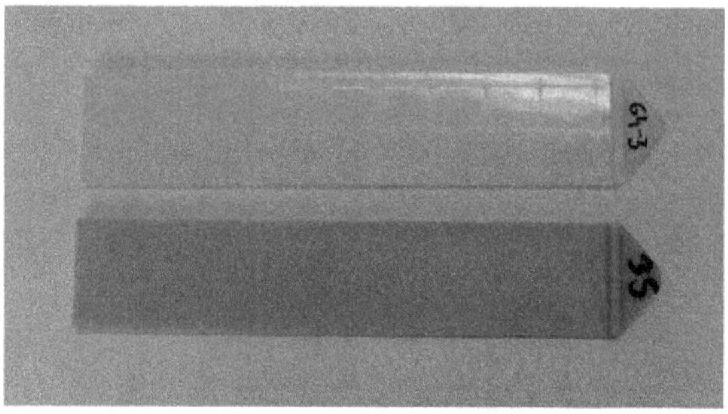

Bonded Semi-Products

Due to the need to save materials, e.g. with special features (including exceptional mechanical, heat-resistant, difficult-to-machine properties or not easily available or expensive) or resulting from machining minimizing (material saving, lower labor and time consumption). Joining of the previously prepared elements can be done in different ways, by welding, soldering and gluing. In many cases, joining elements together allows to obtain a finished product (Figure 5) (Feld M., 2009).

Figure 5. Welded products (above structural steel handle with special friction properties material; below structural steel handle with heat-resistant material)

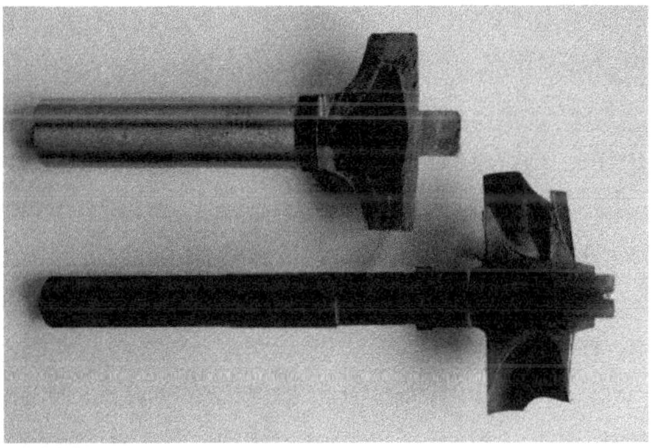

Figure 6. A welded product made of two parts using metallurgical materials (material saving)

Figure 7. Riveted aircraft hull construction

Die Forgings

Forgings are made of a previously prepared material, e.g. a cut-off piece of a rod. Machining involves free upsetting (with a hammer) or die forging (with a hammer press or screw press) of the material to achieve the desired shape of the object. By forging, objects with the shape of a prism, cylinder, of various cross-sections (including irregular or variable), plates, discs, products with or without punched holes, sleeves, and many more are obtained. Due to the fact that machining is hot, the forging surface can be covered with pits and have a considerable roughness. Forgings can be glued to elements cut out of sheet metal, bars, etc., after careful preparation of the surface, both in terms of purity, dimension and shape. Comparing the use of various semi-finished products in adhesive bonding operations, forgings are used much less frequently (Godzimirski J. & Others, 2010; Feld M., 2009; Korzyński M., 2008).

Figure 8. Forgings

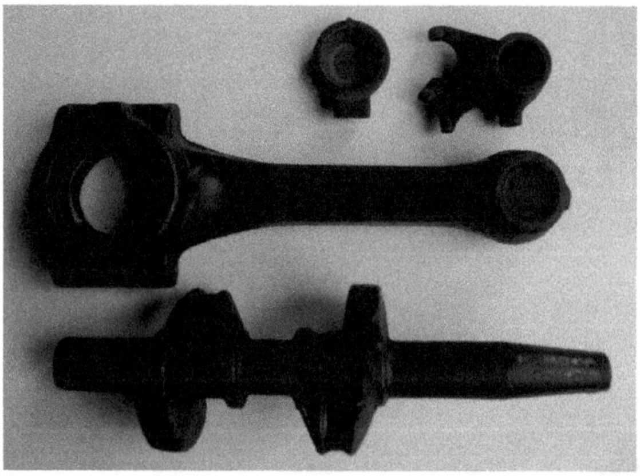

Castings

Semi-finished casting products are obtained by filling the mold with metal or plastic in a liquid or semi-liquid state. Castings are often used in the industry, alongside metallurgical materials, mainly rolled ones (Korzyński, 2008). Casting is used in the manufacturing of products of various shapes, often complicated, difficult or impossible to obtain by other methods. Accuracy of castings may vary a lot, starting from castings made in boxes (often manually), which are the least accurate with surplus material left for further processing, through more accurate castings made in chill molds in a centrifugal method, up to precision castings free from flashes which provide finished products. In order to bond castings together using adhesives, it is important to assess not only the degree of their matching or surface roughness, but also porous areas, characteristic for castings, yet undesirable when adhesive bonding because of adverse stress concentrations, excessive cavities and voids which, if not filled with glue, may affect bonding quality (Korzyński M., 2008; Choroszy B., 2000).

Figure 9. Casting

The group of castings also includes elements made of plastic. In this way, small-size components, such as models and prototypes, as well as machine frames made of polyamide or polyurethane are manufactured. Sometimes the resin cast is reinforced with fibers, nanoparticles or formed by repeatedly pouring thin layers, often with additional elements in the form of fabrics or sheets, obtaining the so-called laminates (Krywult B., 1990).

COMPOSITES

In recent years, many materials, including metal alloys, have been replaced by composites. They are used in space technology, military industry, automotive industry, sports equipment production, products for daily use (Kuczmaszewski J., 2011).

Composite is a material made of at least two components, which has new and usually better properties compared with the components considered separately. The components of the composite do not dissolve or combine with other chemical compounds, yet have an effect on one another.

A typical **composite consists of** a matrix and reinforcement (filler) with better mechanical properties. The matrix has the following functions:

- to protect the reinforcement against mechanical damage,
- to transfer external stresses to the reinforcement,
- to prevent cracks forming,
- to give the products the desired shape.

The purpose of the reinforcement is to strengthen the material and improve its mechanical properties. It is important to create a strong bond between the matrix and the reinforcement.

Composites are divided according to the type of matrix and reinforcement (Table 11).

Table 11. Composites

Types of composites	
Matrix material	**Reinforcement material**
- polymer - polyester, polypropylene, polyurethane, epoxy resins - metal - steel, aluminum - ceramics	- polymer - polyethylene, aramid fibers, nylon, natural polymers - metal - titanium, iron, molybdenum - ceramics - Si3N4, other nitrides, glass fiber, carbides, oxides

Because of the way the reinforcement is made, composites are divided into three groups:

- **Molecular**: strengthening with large particles of another phase (e.g. polymer with filler, concrete), as well as dispersion hardened materials, where reinforcement particles have diameters of 0.01 to 0.1 mm (metal alloys with metal matrix and hard phase reinforcement, e.g. thorium in nickel). Their properties depend on the distribution and geometrical features of the particles. For this reason, particles should have the same dimensions, should be evenly distributed and have an optimal percentage of the volume occupied by particles - (depending on the materials and expected properties of the composite),

- **Fiber Reinforced**: these are the most common materials; most commonly used are fibers (elements with a length much larger than the diameter of 0.001 to 0.15 mm), carbon, aramid (kevlar), glass (Figure 10). The fibers can be continuous (very long, length L/D diameter D>>>100) and discontinuous (L/D=about 100), structured or unstructured. The most popular are composites in which the fibers are reinforced with polymers (polyester, vinyl esters). There are also fibers with metallic and ceramic matrix. The strength of the composite is influenced by the length, diameter and arrangement of the fiber,
- **Structural** (laminates and sandwich layer composites): composed of multiple layers with different fiber orientation. To manufacture these, fabrics instead of unbound fibers are used.

Polymer based composites (resins) reinforced with glass, carbon or aramid fabrics tend to be used. Fabrics may be woven differently, and the layers can be arranged by either orienting them in the same direction (each layer in the same direction) or alternately (e.g. with a different angle of weaving). The laminates are produced manually or in molds (Figure 11).

Manual manufacturing involves laying down the fibers and soaking them in resin, layer by layer. It is a time-consuming method used in the production of small parts. Such elements are formed under pressure or in a vacuum. On an industrial scale, a form is used in which the fibers are placed, filled with resin, heated and alternatively placed in a vacuum.

The sandwich or honeycomb composites, on the other hand, are two strong outer layers separated by a weaker, less dense core material (Figure 12). Such a structure is used in the construction of roofs, walls and airplane wings.

Composites are characterized by unique properties, which cannot be achieved with monolithic materials (Sobieszczański J., 2004; Oczoś K., Kawalec A., 2012):

- may be used to manufacture products of complex shapes and sizes (complete elements of air and space hulls, car trunks) (Figure 13),
- have high stiffness, strength and plasticity,
- have resistance to brittle cracking, fatigue and impact resistance,
- have appropriate density (low in aircraft constructions),
- have resistance to high temperature or cryogenic conditions.

However, composites also have disadvantageous features:

- micro-cracks are formed in them
- the connection between the fibers and the matrix can be damaged
- the damage is difficult to detect and repair
- machining is difficult, therefore other methods of shaping products have to be used: laser processing or adhesive bonding.

Figure 10. Examples of different types of composite matrices: a) carbon fiber (black) b) aramid fiber (yellow), c) glass fiber (white), d) bicolorfiber – diolene (black and blue), e) carbon-fiber hybrid fabric f) glass mat aramid (black and yellow)

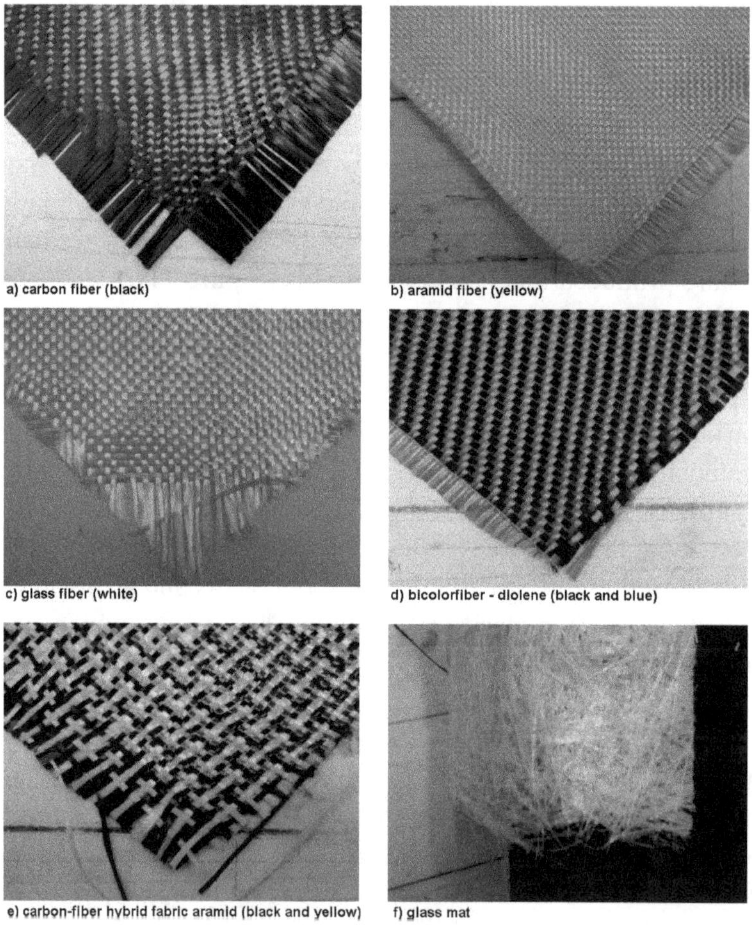

Figure 11. Laminate

Figure 12. Structural material of glued structures: (a) honeycomb structure, (b) disc glued structure

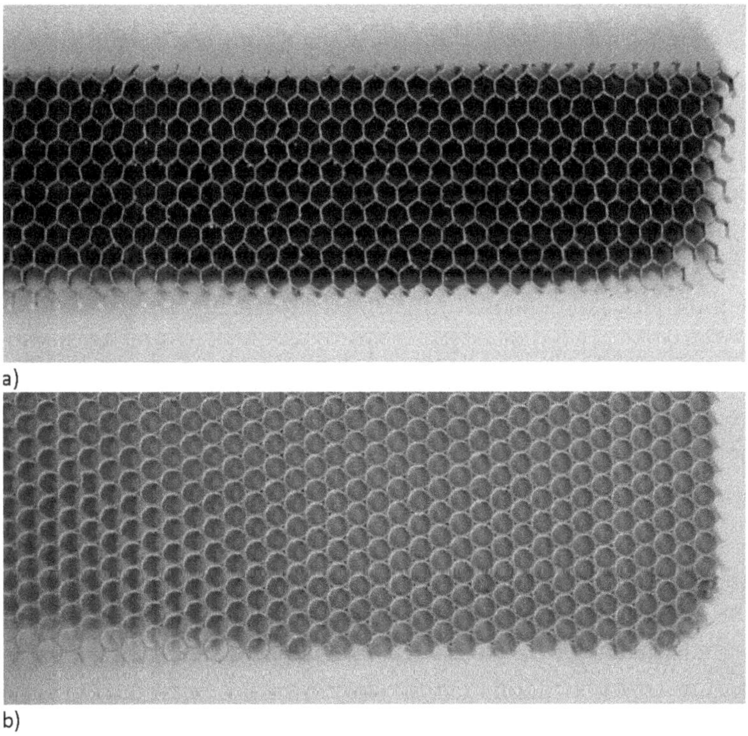

Figure 13. Adhesive products made of composites: a) pilot seat of a glider, b) airplane fuselage fixed together with wings using adhesives

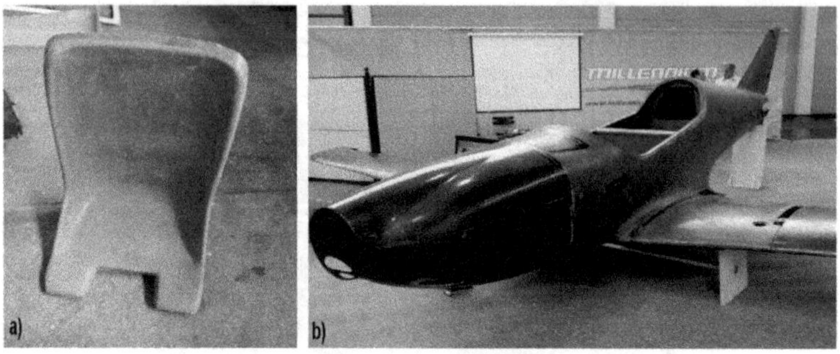

Selection of the starting material in a technological process is not an easy task. There are many factors which should be taken into consideration when selecting it: material grade, properties obtained after initial shaping operations, accuracy of the shape and dimensions, surface roughness, planned machining, and many others. Sometimes, characteristics of a given material exclude or indicate a specific, more favorable way of manufacturing of the finished product. An example would be adhesive-bonded semi-finished products or wood (especially hard) which might be poorly machinable, would split, chip, and have splinters or torn grain. Sometimes, metal alloys (e.g. titanium) form long, stringy chips during cutting, making machining difficult. They get wrapped up around the machined piece and the tool, cause hot parts to stick together, etc. Also, machining elastic materials might pose a risk of loss of rigidity, deformation of the workpiece or vibration. In such case, an alternative method should be considered, e.g. casting, die stamping, or cold pressure welding to make a valuable product (Bielawski R. & Others, 2015; Wolf, 2020).

ADHESIVE SELECTION FOR GLUED MATERIALS

Given a great diversity of adhesives available on the market, there are a number of factors to be taken into consideration when preparing for a gluing process. These are, above all (Kuczmaszewski J., 1995):

- the type of materials to be bonded,
- end use of the product,

- technological capability of the plant, e.g. accuracy and preparation speed of surfaces to be glued, time between glue preparation and its application, available devices for pressing or heating,
- industrial discipline involving dispensing, mixing, glue application, cleanliness,
- future operating conditions of glued products, taking into account temperature, type of load, humidity and chemical composition of the working environment.

When production conditions are known, the adhesive manufacturer's specifications can be used to select the optimal solution. An example of a brief specification is given in the table below (Table 11).

Table 11. The example of commercial information about product properties (Epoxy, 2019)

Type of glue: Loctite 3422 - epoxy	
Shear strength obtained	After 7 days at 23°C
Glued materials	Bonding strength, N/mm^2
Shot blasted steel	18
Stainless steel	5
Electrolytic zinc coating	5,5
Ground aluminum,	4
Pickled aluminum,	9,7
Polycarbonate	1
Rubber-phenolic resin compounds	1
Brass	4
GRP	0,7
ABS	0,7
Hardwood	8
Softwood	8,5

Manufacturers of adhesives also provide information about the materials for which a given adhesive is intended (Table 12).

CONCLUSION

Currently, a number of structural materials with diversified properties are known. Manufacturing techniques which are known and applied allow to shape them in a specific way to create structures useful to man. In order to obtain a finished product with specific characteristics, it is necessary not

Table 12. Application of selected adhesives available comercially (Farnell, 2019; Albeco, 2020)

Type of adhesive	Trade name	Application
Epoxy	Araldite 2011	Metals, hard plastics, rubber, glass
	Araldite 2014	Metals, plastics, glass in structures exposed to chemicals, high temperature
	Araldite 2015	Laminates, GMR-type presses, SMCs, metals
	Multibond 1101	Laminates, metals, ceramics, stone, wood, concrete in heavily loaded structures exposed to temperature, chemicals, water, etc.
	Multibond 15	Metals: structural steel, stainless steel, acid resistant, galvanized; aluminum, laminates, glass, concrete, ceramics, stone
Methacrylate	Araldite 2021-1	Plastics, composites, metals
	Acralock PP1-02	Polyolefins: PE, PP, POM, PTFE
	Loctite 276	Structural steel, stainless steel, brass, nickel-plated surfaces
Polyurethane	Soudal 66A	Wood, concrete, stone, brick, metals
	Multibond 3314	Sandwich laminate, rubber
Cyanoacrylate	3M PR20	Plastics, rubber
	3M PR100	Metals, plastics, rubber

only to select the grade of the material, but also to develop a technological process of its manufacture. Due to the adhesive operations that are often used and analyzed in this study, the selection of a semi-finished product should be considered due to its suitability for such processes. Not all the methods of shaping semi-finished products are equally useful for adhesive bonding. Flat surfaces, e.g. metal sheets, sections, plastic boards, wooden boards or glass will be easier to bond together. On the other hand, surfaces of other forms, such as metal castings or ceramic products can be difficult or even impossible to glue together. The reason for this can be the difficulty of matching the elements to each other or the unfavorable properties of the surface layer, such as roughness or porosity.

REFERENCES

Albeco. (2020). *Structural adhesives Catalog*. Retrieved from https://www.albeco.com.pl/data/catalogueb/Loctite%20-%20Kleje%20Epoksydowe%20Loctite%20Hysol.pdf

Aluminum. (2020). *Aluminium i jego stopy* [Aluminum and its alloys]. Retrieved from https://strefainzyniera.pl/artykul/1234/aluminium-i-jego-stopy

Ataszek. (2019). Retrieved from https://ataszek.pl/273-klej-metakrylowy-do-pp-pe-ptfe-acralock-pp-1-02.html?gclid=EAIaIQobChMI3PD_ioyz5QIVBOd3Ch1DeALhEAQYASABEgLfQvD_BwE

Bernaciak M. (2010). Technologia klejenia w montażu paneli słonecznych [Gluing technology in the installation of solar panels]. *Projektowanie i Konstrukcje Inżynierskie, 3*(30).

Bielawski, R., Rządkowski, W., Augustyn, S., & Pyrzanowski, P. (2015). Nowoczesne materiały stosowane w konstrukcjach lotniczych – wybrane problemy oraz kierunki rozwoju [Modern materials used in aircraft constructions - selected problems and directions for development]. *Rzeszów: Zeszyty Naukowe Politechniki Rzeszowskiej, 87*(3/15), 203-216.

Blicharski, M. (2004). Inżynieria materiałowa. Stal. [Material Engineering. Steel]. Warszawa: Wydawnictwa Naukowo-Techniczne.

Buchnik, G. (2006). *Fascination of sheet metal. A material of limitless possibilities*. Würzburg: Vogel Buchverlag.

Choroszy, B. (2000). *Technologia maszyn* [Machine Technology]. Wrocław: Oficyna Wydawnicza Politechniki Wrocławskiej.

Composites. (2020). *Co to są materiały kompozytowe? Zalety kompozytów* [What are composite materials? Advantages of composites]. Retrieved from http://tuznajdziesz.pl/produkty/artykuly/co-sa-materialy-kompozytowe-zalety-kompozytow-1602/

Dobrzański, L. A. (2001). Zasady doboru materiałów inżynierskich z kartami charakterystyk [Rules for the selection of engineering materials with safety data sheets]. Gliwice: Wydawnictwo Politechniki Śląskiej.

Dobrzański, L. A. (2002). Podstawy nauki o materiałach i metaloznawstwo [Fundamentals of materials science and metal science]. Warszawa: Wydawnictwo Naukowo-Techniczne.

Dobrzański, L. A. (2008). Niemetalowe materiały inżynierskie [Non-metallic engineering materials]. Gliwice: Wydawnictwo Politechniki Śląskiej.

Dydek, Ł., Hryniewicz, T., & Rokosz, K. (2016). *Zastosowanie tytanu i wybranych stopów tytanu w lotnictwie* [The use of titanium and selected titanium alloys in aviation]. Autobusy 8.

Epoxy. (2019). *Kleje epoksydowe mieszane 1:1* [1:1 Mixed Epoxy Adhesives]. Retrieved from http://www.kleje-przemyslowe.pl/r01-31-kleje_epoksydowe_mieszane_1_do_1.htm

Farnell. (2019). https://pl.farnell.com/araldite/araldite-2021-1-50ml/epoxy-adhesive-cartridge-yellow/dp/2676048

Feld, M. (2009). *Podstawy projektowania procesów technologicznych typowych części maszyn* [Basics of designing technological processes of typical machine parts]. Warszawa: Wydawnictwa Naukowo-Techniczne.

Godzimirski, J., Komorek, A., Rośkowicz, M., Smal, T., & Tkaczuk, S. (2010). Tworzywa adhezyjne. Zastosowanie w naprawach sprzętu technicznego [Adhesive materials. Application in technical equipment repairs]. Warszawa: Wydawnictwa Naukowo-Techniczne.

Haws, W. J. (2000). New trends in powder processing beryllium-containing alloys, JOM. *Journal of the Minerals Metals & Materials Society*, *52*(5), 35–37. doi:10.100711837-000-0031-2

Korzyński, M. (2008). Podstawy technologii maszyn [Fundamentals of Machine Technology]. Rzeszów: Oficyna Wydawnicza Politechniki Rzeszowskiej.

Krywult, B. (1990). Tworzywa sztuczne [Plastics]. Rzeszów: Wydawnictwa Uczelniane Politechniki Rzeszowskiej.

Kuczmaczewski, J. (1995) Podstawy konstrukcyjne i technologiczne oceny wytrzymałości adhezyjnych połączeń metali [Constructional and technological foundations for assessing the strength of adhesive metal joints]. Lublin: Wydawnictwa Uczelniane Politechniki Lubelskiej.

Kuczmaszewski J. (2011). *Efektywność wytwarzania elementów lotniczych ze stopów aluminium i magnezu* [The efficiency of manufacturing aviation components from aluminum and magnesium alloys]. Lublin: Komputerowo zintegrowane zarządzanie.

Las. (2020). *Atlas*. Retrieved from http://www.wlin.pl/las/atlas-drzew-i-krzewow/

Łunarski, J., & Banaś, G. (1988). Technologia budowy maszyn [Machine construction technology]. Rzeszów: Wyd. Politechniki Rzeszowskiej.

Materion. (2019). Retrieved from https://materion.com/products/beryllium-products

Oczoś, K. (1988). *Kształtowanie materiałów skoncentrowanymi strumieniami energii [Shaping materials with concentrated energy streams].* Rzeszów: Wyd. Politechniki Rzeszowskiej.

Oczoś, K., & Kawalec, A. (2012). Kształtowanie metali lekkich [Light metal shaping]. Warszawa: PWN.

Plast. (2019). Retrieved from https://www.plastech.pl/plastechopedia/UP-UP-R-87

Przybyłowicz, K. (2012). Nowoczesne metaloznawstwo [Modern metal science]. Kraków: Wydawnictwo Naukowe Akapit.

Rudnik, S. (1996). *Metaloznawstwo* [Metallography]. Warszawa: Wydawnictwo Naukowe PWN.

Sobieszczański, J. (2004). *Spajanie* [Joining]. Warszawa: Oficyna Wydawnicza Politechniki Warszawskiej.

Virgamet. (2019). Retrieved from https://virgamet.pl/1-4305-x8crnis189-1-4306-x2crni1911-aisi-304-3041-303-stal-nierdzewna

Westegen H. (1993). Magnesium alloys for structural applications; recent advances. *Journal de Physique IV, 3.*

Wodecki, J. (2011). Podstawy projektowania procesów technologicznych części maszyn [Basics of designing technological processes of machine parts]. Gliwice: Wydawnictwo Politechniki Śląskiej.

Wolf. (2020). *Tytan* [Titanium]. Retrieved from https://www.wolften.pl/dane-techniczne/18

Żuchowska, D. (2000). Polimery konstrukcyjne [Constructional polymers]. Warszawa: Wydawnictwo Naukowo-Techniczne.

Chapter 3
The Importance of the Surface Preparation Method

ABSTRACT

In this chapter, the specific steps of the bonding operation were presented. Problems related to the preparation of surfaces to be glued together concerning different construction materials were described. In addition, attention was paid to the issues concerning the preparation of the adhesive mass, techniques of glue application, and the requirements necessary to be met during the mass curing time. The applied equipment necessary to obtain a durable adhesive bond was characterized. Due to many factors influencing the effects of gluing, the methods of assessing the strength of joints used in the industry were presented. The second part of the chapter analyzes the importance of preparing the material surface before applying the adhesive, the phenomenon of adhesion and cohesion, and the problem of ensuring proper wettability. Surface free energy determination issues and the mechanical theory of adhesion were also described in this chapter.

ADHESIVE BONDING PROCEDURE

Adhesive bonding technology is a multistage process, the complexity of which depends on, among others, the type of bonded materials, the selected adhesive, available workshop equipment and the shape of bonded surfaces.

DOI: 10.4018/978-1-7998-4634-5.ch003

Adhesive bonding takes place according to strictly defined rules, with maintaining the order of performing the consecutive stages (Porębska M., Skorupa A, 1997; Szabajkowicz W., Towt A., 1994):

- preparing the surfaces of objects for adhesive bonding,
- preparing the resin compound,
- bonding and pressing,
- drying, curing, ageing,
- finishing processes,
- bonding quality control.

Errors in performing or omission of a particular stage may result in obtaining a bond of improper properties, e.g. too low mechanical strength, insufficient resistance to ambient conditions, chemicals, insufficient aesthetic values, etc. (Brockmann W. & Others, 2009).

Surface Preparation

Preparation of the surface before bonding is a key stage of the bonding process. Initially the surface is controlled visually, rejecting defective elements, the ones that include pits, deep or vast corrosion centers (difficult to remove), or which are cracked. Then the following activities are performed and they are aimed at (Porębska M., Skorupa A, 1997; Wiącek Z., 1987):

- removal of impurities – dust, dirt, oil, remaining paints, insignificant and shallow corrosion centers, oxide layers,
- proper surface development – in order to acquire particular surface roughness and stereometric features, i.e. non-directional structure of particular height of elevations and pits,
- surface activation – improvement of adhesion conditions (acquiring better wettability).

Surface preparation can be performed with chemical or mechanical methods. The choice of a method depends first and foremost on the type of the bonded material. Thus, the components made of (Puff T., Sołtys W., 1980; Rudawska A., Kuczmaszewski J., 2005; Szabajkowicz W., Towt A., 1994; Kuczmaszewski J.,1990; Rudawska A., Chruściel M., 2011):

- aluminum and aluminum alloys - are chemically prepared (acid or alkaline etching), electro-chemical (anodizing), or by means of grinding, abrasive blasting,
- steel – is prepared in a more simplified way, compared to aluminum and aluminum alloys; it results from susceptibility of steel to corrosion (a layer of detrimental oxides), which decreases the component's effective area; machining is applied (grinding, sand-blasting), and then degreasing is performed in liquid trichloroethylene or its vapors, petrol, or acetone,
- copper and copper alloys – are prepared mechanically- in the process of sand-blasting or by using a wire brush (after the process there occurs degreasing), or chemically – in a concentrated ammonia bath, followed by flushing with water and drying,
- glass and ceramics – are usually only degreased,
- plastics – are rubbed until obtaining particular roughness and are then dust extracted (PTFE, however, is a plastic which is very difficult for adhesive bonding and it requires making the surface effective by means of a solution of metallic sodium in liquid ammonia, or a solution of sodium in naphthalene).

Application of a particular mechanical or chemical method may require additional processes, such as degreasing (e.g. in acetone, kerosene, TRI), rinsing in water and drying, and sometimes applying primers (substances that facilitate curing) before adhesive is applied.

Preparing the Resin Compound

Depending on the selected adhesive type, a particular portion of homogenous mass of a particular consistence is prepared. One-component adhesives do not pose any significant problems, as they are squeezed out of a box manually or mechanically, with or without heating the component or adhesive.

The mass of multi-component adhesives needs to be carefully applied by measuring the appropriate amount of all components with the use of separate spatulas, applicators, or by means of proper stirrers, whose role is to stir the mass until a uniform color and consistence are obtained. The sequence of adding ingredients should also be respected. Firstly, additives that provide viscosity are added to the adhesive substance, followed by additives that regulate curing speed, thermal resistance, after-tack, hardening agent, and finally, the filler (Szabajkowicz W., Towt A., 1994).

Figure 1. Surface cleaning before adhesive bonding: a) in a sanding machine for sand blasting with corundum, b) with coated tools, c) abrasive paper, d) chemical, e) in ultrasonic cleaner

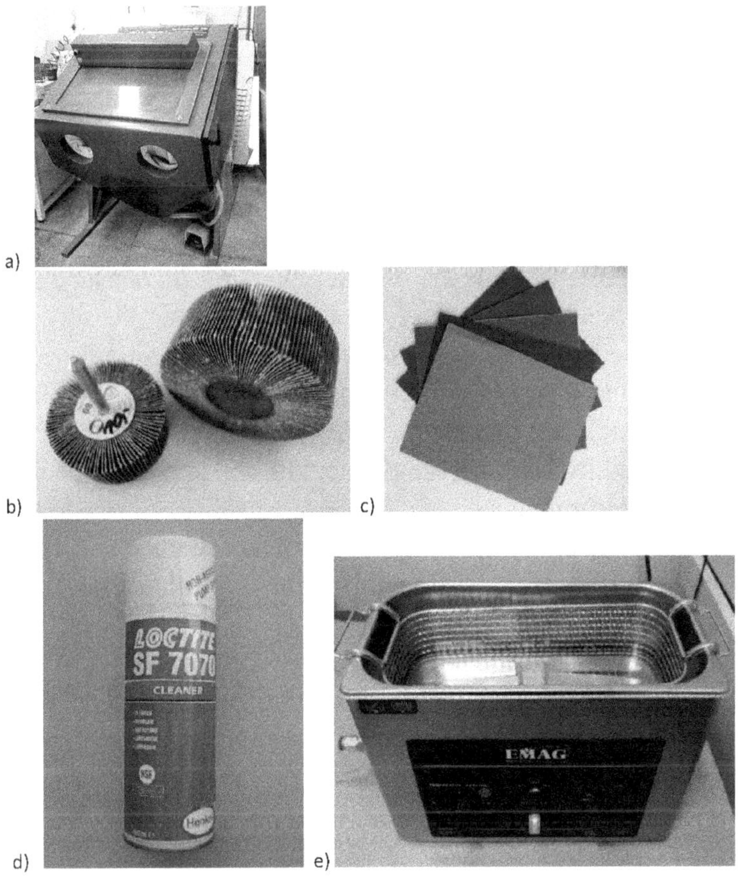

While preparing the resin compound, appropriate ambient conditions should be provided: cleanliness of air, temperature and humidity. Sometimes it is necessary to heat the mass or to expel any air.

The process of stirring adhesive ingredients may be carried out with the use of:

- manual or mechanic stirrers,
- stirring applicators,
- dispenser guns – manual or mechanic.

Adhesive Application

The prepared adhesive mass should be applied immediately. Depending on the form, adhesives may be applied on the surface in different ways by using additional equipment (Puff T., Sołtys W., 1980; Rudawska A., 2016; Szabajkowicz W., Towt A., 1994):

- liquid adhesives - liquids applied by a brush or by using a spray method with a dispenser gun, as well as with rolls damped with an adhesive in a continuous manner, by means of dipping,
- polishes – applied by means of a scraper or a dispenser gun through rubbing,
- films – applied by rubbing, without any additional tools, yet often on the surface that was additionally spread with an adhesive (in that case application tools are used),
- bars – applied by rubbing the part being bonded with special heating rods, dispenser guns;
- powders – applied by sprinkling the surface with a sprayer and then heating it until the powder transforms into a liquid state.

In the case of mechanized and automated processes, a layer of unitary and repetitive thickness is obtained. However, increased adhesive consumption should be anticipated.

Figure 2. Application of the adhesive: a) with a brush, b) with a spatula, c) with a hot gun

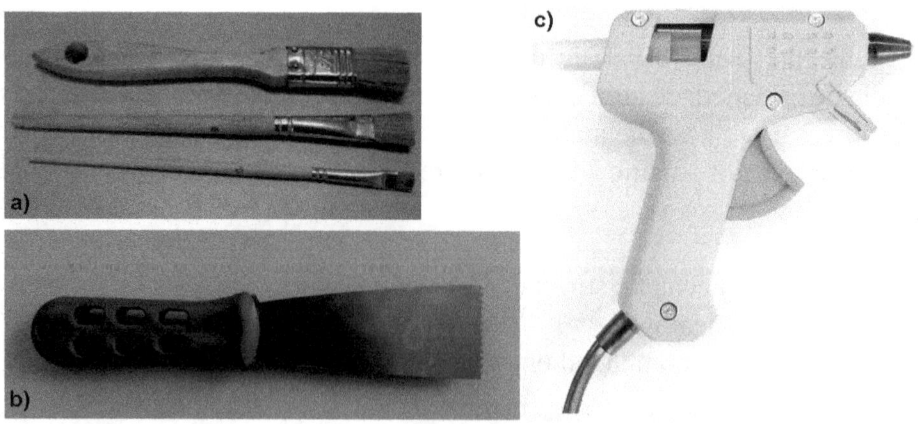

Bonding, Pressing, Curing

After applying the adhesive layer the bonded elements are assembled and pressed. In order to provide dimensional repetitiveness and permanence of components' location, special handles with pressing elements are constructed.

Especially in the case of cold gluing, cold pressing is essential due to the fact that the cured adhesive increases its volume (volatile fractions are released). Pressing may also ensure that a particular and homogenous joint thickness is obtained, as well as enable appropriate positioning of elements. Therefore, handles made especially for adhesive bonding, equipped with stirrup bolts, spring clips, pressure beams, presses or pneumatic devices are very common. Vacuum or pressure chambers are also applied. The choice of pressing force (pressure) depends on the type of adhesive, especially its consistence, viscosity and density. Exerting pressure should take place immediately after assembling the components to avoid early joint curing. Curing may take place without contact of adhesive with the component's surface and it is an irreversible process. In many cases, the pressing force is set experimentally, however it is sometimes recommended to use for adhesives of small curing contraction <0.1 MPa, or for adhesives of big curing contraction of approximately 05-2.0 MPa(Godzimirski J, & Others, 1997; Rudawska A., Kuczmaszewski J., 2005) .

The pressing force determines the cohesive strength of the joint.

Figure 3. Example of a special adhesive holder for laboratory applications

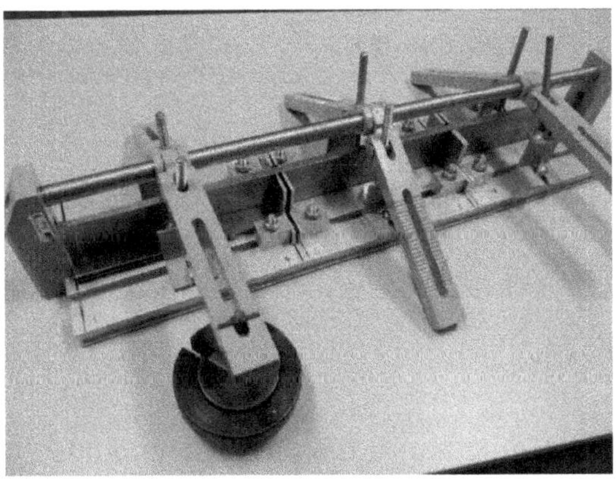

When thin-wall material having low stiffness are glued, the locking fasteners and fixtures should not be used locally, but rather comprise an additional point of support, which enables even distribution of the pressing force (of a larger area, flat or fitted to the component in terms of shape, most often cylindrical) (Rudawska A., 2016).

Sometimes before the surface is glued, it is necessary to remove the excess of the solvent by vaporization. Then, before eventual pressing of elements to each other, an already applied layer of adhesive is pre-dried. It is justified by the necessity to acquire adequate strength. Due to a large amount of solvent, there is a possibility that some of its part will stay in the joint and become a point of internal stresses and, thereby lower the joint's strength. The drying time is chosen with consideration of the adhesive type, volatility of the applied solvent, air temperature and humidity, air circulation etc. (in reality, the drying time shall range from 1 to several dozen of hours) (Puff T., Sołtys W., 1980).

The selection of curing parameters depends on the selected adhesive and chemical reactions taking place in the joint. For each adhesive composition temperature, time and pressure that are proper to bond the element are determined. In most cases manufacturers define the parameters and provide them in the specification.

Example 1 (Loctite, 2019):

For LOCTITE 3422 adhesive:

- hardening velocity, depending on time/temperature – mixed in the ratio of 1:1, in terms of volume or mass, reaches, in room temperature, high strength in 2 hours,
- glued parts are to be fixed (initial strength: 0.1 N/m^2) after 7 minutes in room temperature and a gap of 0,05 mm. In order to accelerate hardening higher temperatures may be applied.
  Example 2 (Farnell, 2019):

For Araldite 2021 adhesive:

- curing in room temperature, the adhesive includes surface activators (primer),
- cures evenly in the whole mass in 5 minutes,
- there is a possibility to cure in low temperatures (even in 10°C).

Figure 4. Example of a primer with adhesive (primer is used separately and is applied to the prepared surface before application of the glue)

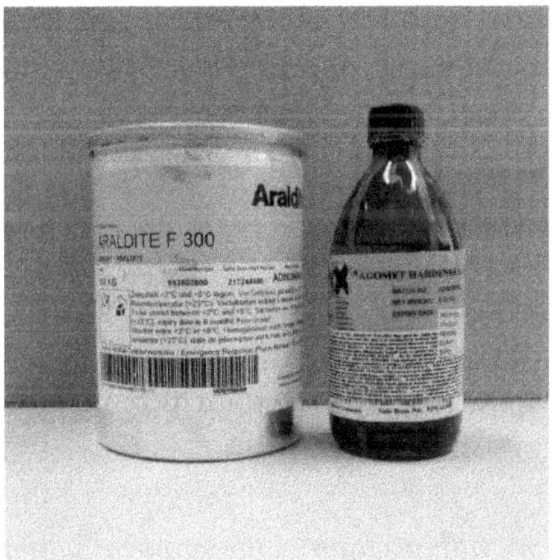

In some circumstances, apart from the tools used to acquire proper pressure, heating devices are also necessary.

When hot gluing is applied, electric radiators or infrared radiators with appropriate controllers (to regulate and control) are used in order to heat and maintain the proper temperature. Heat presses are used, with the view to obtain the simultaneous effect of pressing and increased temperature. However, difficulties may be encountered in making an accurate assessment of the size of the pressure and its control before heating is initiated. Also, the possibility of squeezing the heated adhesive and instability of the pressing due to the decrease of adhesive volume should be taken into account. Therefore, the most effective adhesive-bonding effects are obtained in autoclaves (air-pressure devices equipped with electric radiators), although they limit, in terms of dimensions, chambers or sizes of other elements (Rudawska A., 2016). In the case of the necessity of heat pressing, heating the joint causes excessively dense curing, and may thus lead to excessive brittleness of the joint or damage to the adhesive structure (Czaplicki J. & Others, 1987).

Many modern types of glue (e.g. epoxy ones) do not require either significant pressing, or increased temperature. In such a case, only the requirement of stable location of elements in relation to each other (protection from moving) has to be met.

Sometimes the hardening procedure is neglected; this might be the case of temporary repairs, prototypes, temporary modernization when obtaining target strength is not the priority (Porębska M., Skorupa A, 1997).

Finishing Processes

Theoretically, a well-made joint does not require further processing. In some cases, due to possible squeezing of adhesive while curing, it is necessary to remove scrap, which is not considered to be very aesthetic. This is best to be performed even before the ultimate glue hardening, by using a cloth soaked in a solvent that is appropriate for the glue. In turn, when the finishing operation is planned after curing of the glue, manual grinding with the use of grinders or file blades is applied. Prior to this, however, it might be beneficial to use sponges soaked in a solvent or, as in the case of epoxy resins, in formic acid in order to soften the hard scrap (Czaplicki J. & Others, 1987).

Figure 5. Example of a joint that requires the removal of a mold (a), manual filing file (b)

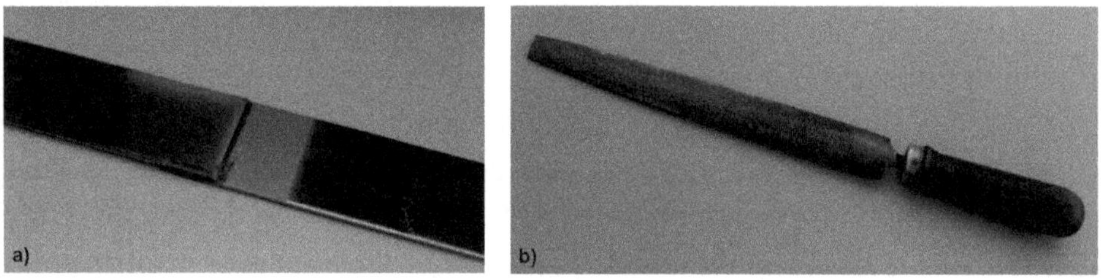

Adhesive Bonding Strength Evaluation

After the elements are bonded and permanent joint connections are obtained, they are subjected to quality control. It can be destructive or non-destructive, depending on whether or not the defects will occur. As a result of joints, the following imperfections may be found, among others: non-bonded spots, joint thickenings, incomplete curing of the adhesive, air bubbles in the joint, cracks, foreign objects, adhesive charing, and sometimes also corrosion of the edges or mechanical damage (Fraunhofer, 2015).

Non-destructive testing takes place without disjoining of the bonded materials; it can be performed in a number of ways:

- review of production procedures – checking real conditions of adhesion with technological assumptions,

- fitting control of the bonded elements, edge location assessment, location according to axes, etc.,
- joint visual control – paying attention to scraps, ill bondages, adhesive layer thickness, etc.;
- joint quality control, e.g. with the Roentgen, acoustic, holographic, thermographic, magnetic methods in order to determine the presence of undesirable impurities, porosity, gas bubbles or tightness (Czaplicki J. & Others, 1987; Godzimirski J., 2009; Szabajkowicz W., Towt A., 1994).

Destructive testing consists in destroying the joint in a particular way and gaining information on specific features of the joint. In manufacturing, destructive tests are performed in the form of sample tests taken from each production batch, as well as in the form of operation tests. Usually, they are called strength tests, because shear strength, peel strength, cleavage strength, fatigue strength are mainly measured and aging tests and tests of resistance to various external conditions are performed (Godzimirski J. & Others, 2010; Porębska M., Skorupa A, 1997; Karny M., 2016).

Shear strength concerns most often lap joints, often occurring in machine constructions. The shear strength tests consist in applying the force acting parallel to the bonded surface. The ultimate test result is influenced by joint thickness and dimensions of the lap (Grant L. D. R. & Others 2009).

Peel strength in butt joints (large-scale) that are loaded with tensile strength in the joint axis (Zielecki W., 2005).

Cleavage strength is tested in such a way that one of the detached parts is static and the second one is ripped off; the force acts almost perpendicularly to the bonding surface (it is the method of detaching rolls). The sample thickness influences the result.

Fatigue strength is determined due to the fact that because of cyclic or non-cyclic variable loads in joints, the process of changes called fatigue takes place. It causes a decrease in durability and hardness of joints, and its final result is joint fracture. The practice shows that fatigue destruction occurs suddenly and with tensions much lower than in the case of joint static strength. Fatigue strength is tested at special stations, where it is possible to apply variable, cyclic load of particular frequency on the joint and to determine the number of cycles, where joint destruction occurred (Godzimirski J, & Others, 1997).

Aging resistance and resistance to variable external conditions (climate or environmental, short or long term studies) in natural conditions or conditions that simulate operation conditions these are, in some cases, accelerated tests

Figure 6. Static shear strength tests (a), visualization of the test (b)

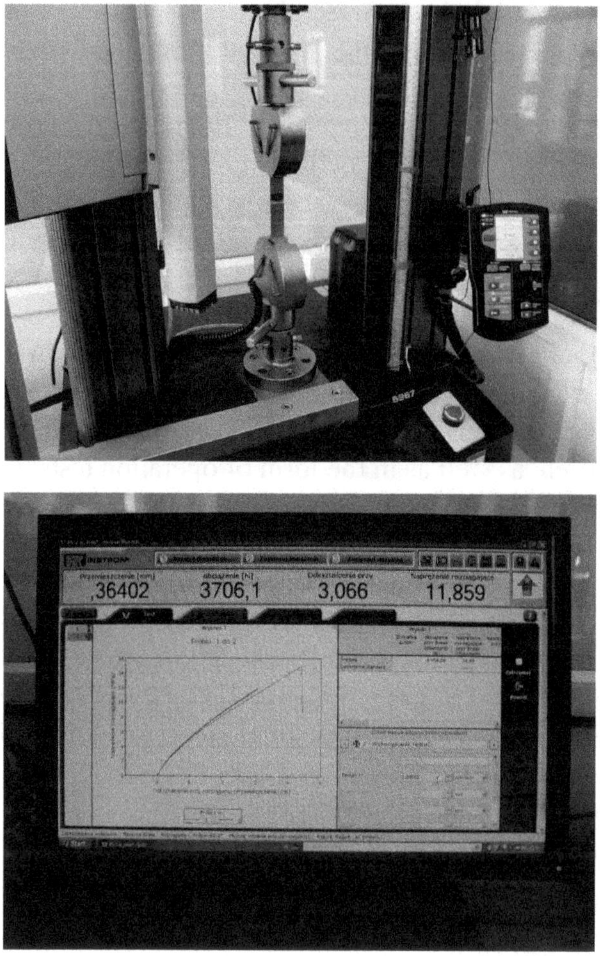

a)

b)

carried out in special chambers of variable and regulated temperature, humidity and pressure. After some time and performing a given number of cycles of changing conditions (dynamic tests), samples are taken out and tested for shear strength. Such analyses are especially desirable in the automotive, shipbuilding and aviation industry (Deb A. & Others, 2007). A variation of aging tests is a test carried out in a salt spray chamber, in which salt spray interacts with the placed samples; in this way the influence of corrosion environment is simulated (Muller M., 2010). Also, tests of resistance to the influence of fuels, greases, exhaust gases are performed in the chambers. Sometimes the strength of joints susceptible to untypical conditions is tested; these include

CO_2-reduced air or other chemical substances, such as fertilizers or cow dung in agriculture (Gaukler J. C., 2009; Muller M., 2010). Some chambers enable placing the joined elements and not only maintaining them in certain external conditions, but also loading them by means of special equipment. In this way, for example, untypical connections due to joint construction in automotive industry are tested (Grant L. D. R. & Others 2009).

Figure 7. Climate test chamber

Thermal shock resistance test is performed in a special chamber, inside of which it is possible to change the temperature rapidly. The chamber control system not only causes a sudden change in temperature, but also provides a series of its changes, depending on the needs. In such a case the so-called thermal shock resistance is determined with the scale and character of a possible loss of operational features (e.g. loss of adhesive properties of glue and the material, sudden brittleness, lack of tightness etc.) on the edges and inside the sample. Such studies have application in the production of adhesively bonded household products, electronics, metrology, aviation, space exploration, cryogenics and medicine (Dziubak C., Rećko W., 2009).

It is necessary to conduct non-destructive tests as the strength of adhesive bonds is difficult to predict, since it depends on a number of factors (Godzimirski J., 2009; Rudawska A., Dębski H., 2011):

- material: adhesive and cohesion properties, adhesive modulus of elasticity, properties of the glued materials, including the modulus of elasticity, adhesive properties of the surface layer,
- technological: the method of preparing the surface for gluing, the preparation method and applying the resin compound, as well as the conditions for curing, i.e. temperature, pressure and time,
- construction: the method of pressing the bond, bond dimensions, symmetry and the shapes of bond elements,
- operational: due to a variety of real operation conditions, variability of loads, the influence of temperature, chemicals, gases and time. Some of those factors may be predicted, however others function randomly.

IMPORTANCE OF THE SURFACE PREPARATION STAGE

The stage of surface preparation of the components to be bonded is of paramount importance. The strength of adhesive bonds depends primarily on the proper adhesion between the glue and the material, as well as the cohesion of the glue and bonded materials, where wettability of the bonded surfaces is a fundamental condition for a proper bond. All impurities, such as grease, water and metal oxides on the surfaces of the adhesively bonded materials decrease wettability of those surfaces. Covering the surface with any impurity prevents its direct contact with the adhesive. Therefore, there is no humidification of the surface by the adhesive in contaminated areas. This leads to uneven adhesion of the material and the glue, and hence the joint's strength is decreased. In the view of the above, preparation of the surface of a component has, in addition to the choice of an appropriate adhesive, a major influence on the strength of adhesive bond. Thus, element surfaces should be free from foreign matter, homogenous and fully developed (Puff T., Sołtys W., 1980).

Adhesion and Cohesion

Two basic conditions need to be met in order to obtain durable and functional adhesive bonds: proper curing of the bond with the material's surface (adhesion) and reaching proper strength in the adhesive bond (cohesion). In a properly designed and constructed adhesive bond the adhesion forces should be approximately equal to the value of the cohesion force. An illustration of the phenomena responsible for creation of adhesive bonds is shown in the drawing (Rudawska A., Jacniacka E., 2009):

Figure 8. Adhesion and cohesion

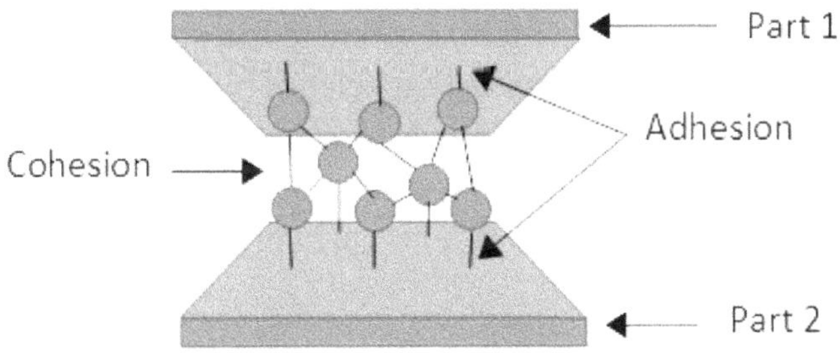

Adhesion (lat. 'adhesio' – joining, adhering), is the phenomenon of surface curing of surface layers of two components, usually a liquid and a solid one.

Cohesion (lat. 'coheasio'), is a property of objects that involves resistance to disintegration of their parts. In the case of the glue, cohesion results in tenacity of the glue layer. The term 'cohesion' is connected with decohesion, which is the effect of forces that are opposite to interatomic and intermolecular interactions and cause object disintegration – disintegration of the curing.

The division of adhesion into specific and mechanic is distinguished.

Specific Adhesion

The basic theory of intermolecular interactions, also known as physiochemical theory, explains in an exhaustive manner the phenomenon of adhesion, treating it as a result of van der Waals intermolecular forces, dispersive forces, dipole or induction forces that occur between glue particles and bonded materials within the range of one micrometer (Przygocki W., Włochowicz A., 2011).

Van der Waals forces are rather small, yet play a key role in adhesion processes. It results from their universal character as they can exist between any particles and then their combined interaction is huge. Additionally, they act over significantly larger distances. As compared to forces of other bonds, the bond length r_0 amounts 0,36-50 nm. Thus, the attraction forces do not require particles to move closer to one another (Żenkiewicz M., 2000).

Adhesion has a significant practical meaning, as adhesiveness serves as the basis not only for the gluing technology, but also for applying paint coats, dying fabrics, making insulation layers in electronics, printing techniques (Rudawska A., Jacniacka E., 2009).

In the case of adhesive bonds, the 'solid-liquid-solid' arrangement is examined, which gradually transforms into the 'solid-solid-solid' arrangement. The glue, which initially is in a liquid state, transforms into a cure – a solid in the process of curing and hardening (Żenkiewicz M., 2000).

The bigger the interaction between particles of two components, the bigger the adhesion. This is connected with their actual contact area. In adhesive bonding the increase of this area may be achieved by exerting pressure on the bond, or by adding surface-active agents (primers) (Dutkiewicz E., 1998).

Wetting and Contact Angle

A drop of adhesive, or a drop of any other liquid, may spill over on the surface intended for adhesive bonding to a greater or lesser degree. The spill-over degree of glue depends on intermolecular forces. Glue spills over on the surface even more if particles of the liquid are more attracted by particles of the glued material. In such a case, the glue wets the solid object surface. If the attraction decreases, the glue does not spill over, does not wet the surface (a drop takes an unchangeable, oval shape). If the attraction is very strong, the liquid spills over the surface forming an even layer. In turn, In the indirect case, the drop takes the shape of a flattened hemisphere (Dutkiewicz E., 1998; Przygocki W., Włochowicz A., 2011).

Contact angle is the measure of wettability - the θ angle between the wetted surface of a solid object and the tangential to the meniscus curve of the wetting liquid, derived from the contact point of liquid with solid object surface (Hebda M., 2007).

The value of liquid contact angle placed on the solid state object surface depends on the balance of (Hebda M., 2007):

- cohesion forces of attraction between liquid particles,
- attraction forces of liquid particles in the liquid-solid contact points,
- attraction forces of particles that do not touch the surface, but which are in the range of intermolecular interactions,
- gravitation forces.

Figure 9. View of the droplet applied to the surface (a), contact angle 1 - material surface, 2 - embedded droplet (b)

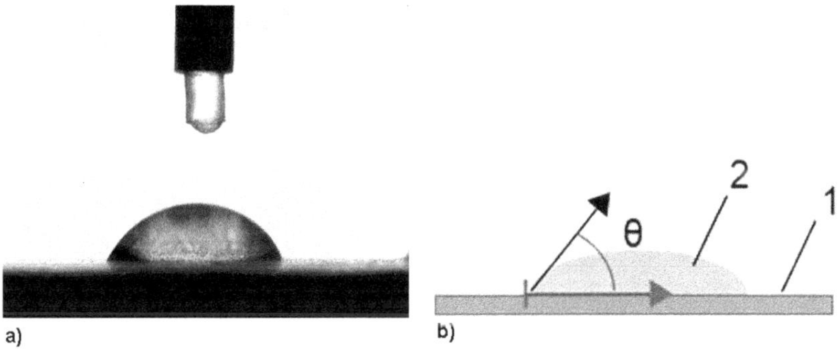

Figure 10. Wettability and its dependence on the contact angle

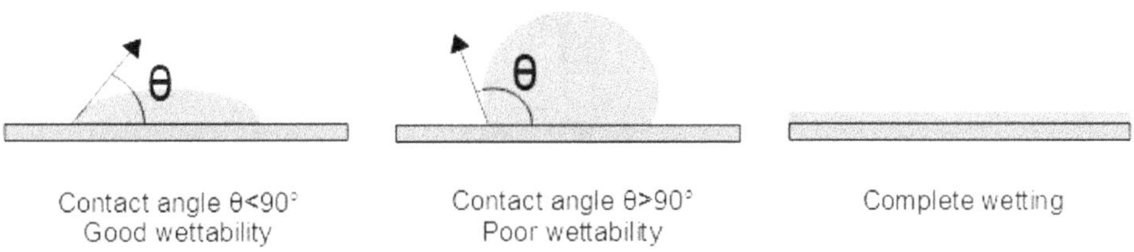

Contact Angle Measurements

A numerous methods exist to measure surface and determine contact angle – pendant drop method, bubble pressure method, strain extensometer method. However, both in laboratory and industrial conditions cameras are used to observe the phenomenon of drop slopping (Przygocki W., Włochowicz A., 2011). Goniometers are used in direct measurements of contact angle - they are devices that allow to read the value of contact angle on the basis of an image of the actual drop of measurement liquid placed on the studied surface. The drop view is visible as smooth object surface has a significant coefficient of light reflection, which provides good drop illumination. Modern goniometers are equipped with additional lighting, which allows to investigate wettability of different solid objects (Przygocki W., Włochowicz A., 2011) (Figure 11).

Figure 11. Contact angle measurements: a) goniometer, b) samples with applied drops of measuring liquids (b) the use of a computer program for visualization of measurements

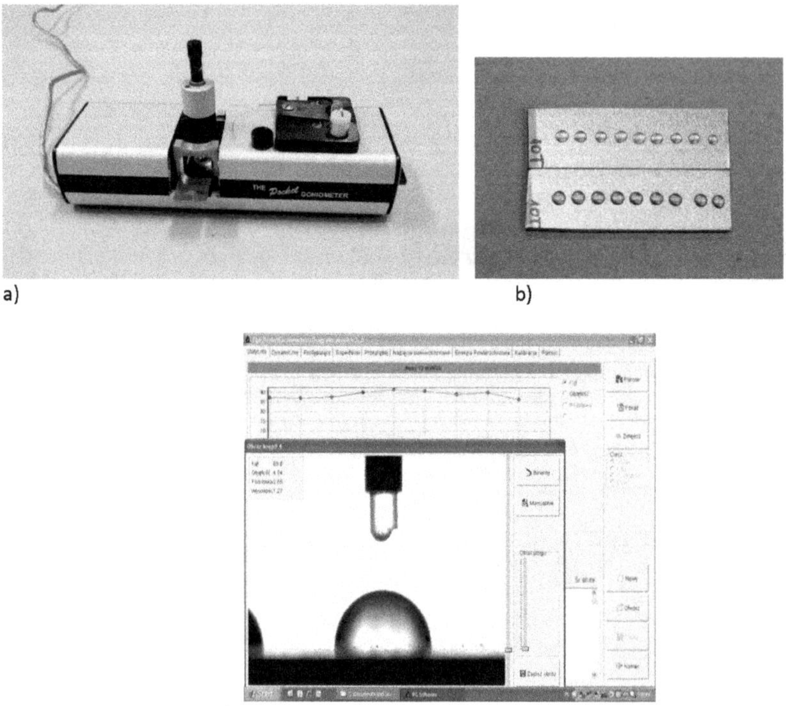

a) b) c)

Surface Free Energy

Surface free energy (SFE) is numerically equal to the work needed to create a new unit of area, during division of two phases that are in balance, in a reversible isothermal process. Its measure, often used to evaluate layer condition of the material's surface, is mJ/m^2. In the process of creating these two areas, two stages are determined. In the first one, directly after the creation of this surface, the atoms maintain their location, whereas in the second one, atoms re-position their location on their own so as to achieve a state of permanent balance. In liquids, these stages occur simultaneously. Thus, in a situation of glue application, the new area is the contact area of both the glue and the object (Żenkiewicz M., 2000).

Surface free energy is a known standard for evaluation of surface preparation for operations that make use of the phenomenon of adhesion: gluing, printing, painting, laminating and metallizing. There are a number of methods for determining SFE of liquids, however it is determined indirectly for solids. An often applied method is the method of contact angle measurements with different liquids (called measure liquid), and calculating on this basis the SFE. The basis of calculation is Young equation, derived from the condition of the balance of forces, which represent relevant surface tension in the point of contact of three phases: solid state, liquid and vapor.

Taking into consideration simplifications (described in references), surface free energy is expressed by the equation (Kłonica M., Kuczmaszewski J., 2011) (Figure 12):

$$\sigma_{SV} = \sigma_{SL} + \sigma_{LV} \bullet \cos\theta$$

where:

σ_{SV} – surface tension on the boundary of solid state-gas state phases
σ_{SL} - surface tension on the boundary of solid state-liquid state phases
σ_{LV} - surface tension on the boundary of liquid state-gas state phases
θ – contact angle

Figure 12. Energy balance illustration

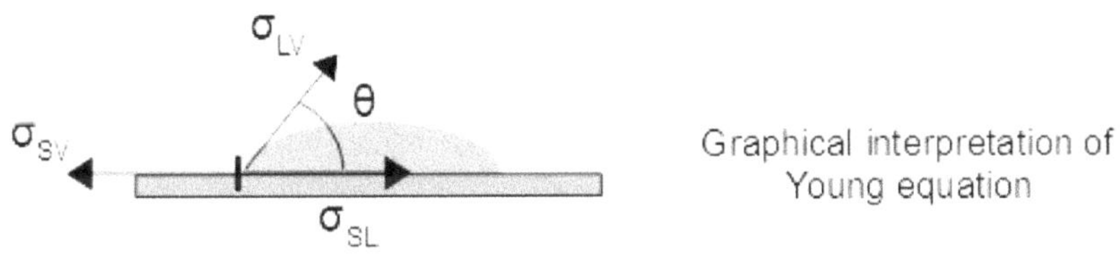

From the energy balance for the quintuple point of three phases it results that:

$$\gamma_{SV} = \gamma_{SL} + \gamma_{LV} \bullet \cos\theta$$

where γ is surface free energy.

The Owens-Wendt method, very often applied in determining SFE, states that surface free energy is a sum of two components: polar and dispersive:

$$\gamma_S = \gamma_S^{p} + \gamma_S^{d}$$

where:

γ_S – surface free energy (SFE),
γ_S^{p}– SFE polar component,
γ_S^{d}– SFE dispersive component.

It is assumed that SFE of the measurement liquid γ_L, as well as the solid object γ_S, also consists of two components (γ_L^{p} and γ_L^{d}). Then, SFE may be determined as:

$$\gamma_L(1 + \cos\theta/2) = (\gamma_S^{p}\gamma_L^{p})^{0,5} + (\gamma_S^{d}\gamma_L^{d})^{0,5}$$

where: θ – contact angle of solid object surface when applying given liquid.

These components being known, it is possible to calculate surface energy of the studied materials $_S$ from the system of equations:

$$(\gamma_L^{d})^{0,5} + 1,53\ (\gamma_S^{p})^{0,5} = 7,8\ (1 + \cos\theta_1)$$

$$(\gamma_L^{d})^{0,5} + 0,22\ (\gamma_S^{p})^{0,5} = 3,65\ (1 + \cos\theta_2)$$

The values of polar and dispersive components of measure liquids may be found in (Żenkiewicz M., 2000), whereby it is recommended that in a measurement with two liquids, one of them should be polar, and the second one non-polar.

Table 1. Values of polar and dispersive components of typical measure liquids (Żenkiewicz M., 2000)

Measurement liquid*	**Liquid surface energy γ_L, mJ/m^2**	**Polar component γ_L^{p}, mJ/m^2**	**Dispersive component γ_L^{d}, mJ/m^2**
Glycerine (p)	64.0	30.0	34.0
Water (p)	72.8	51.0	21.8
Formamide (p)	58.0	19.0	39.0
Methylene iodide (n)	50.8	2.3	48.5
α-bromonaphthalene (n)	44.4	0.9	43.5
(p) – polar liquid (n) – nonpolar liquid			

The abovementioned goniometer is therefore used together with a PC, equipped with a program enabling visualization of measurements, reading the contact angle and making additional calculations of surface free energy with a selected method. In many goniometers it is possible to do measurements with different liquids. In laboratory and industrial practice one (most often water) or two liquids of varied polar and dispersive components are relatively frequently used, e.g. distilled water and methylene iodide. This choice is often determined by software computation capacities (Figure 13).

Figure 13. Use of computer station for wettability measurements (a) and SFE calculations (b)

a)

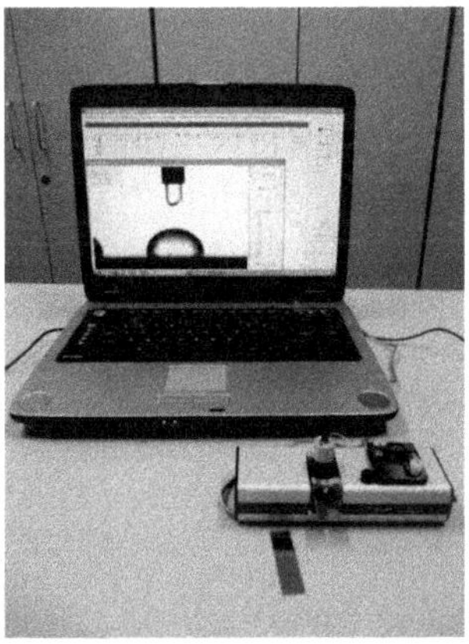

b)

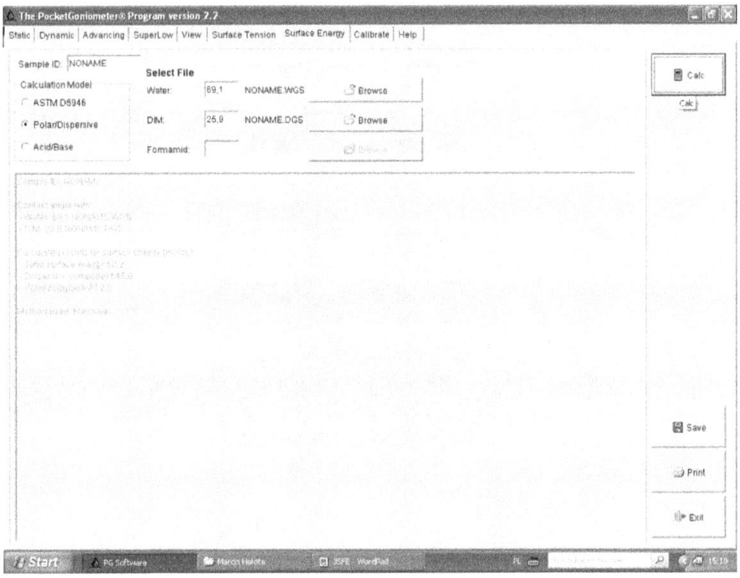

Contact angle measurements that are necessary to determine surface free energy need to be done at a place that is not liable to vibrations, as they may affect the process of slopping of liquid drops. The test bench should be levelled and the room needs to have fixed conditions (temperature and humidity). The investigated surface should be rigid, non-movable, and have appropriate roughness. A condition for correct measurement is also cleanliness (lack of contamination, lubrication substances, liquids), as well as homogeneity. On the other hand, when selecting measurement liquid, one should pay attention that it should not get into chemical interaction with the surface (Żenkiewicz M., 2000).

Mechanical Theory of Adhesion

One of the adhesion theories, apart from the energy theory, is the mechanical theory. According to this theory, the strength of adhesive bonds increases by increasing roughness of the surface of the bonded materials. It was stated that greater roughness of surface translates to a greater number and depth of pits of irregular shapes. These pits can be penetrated by glue, creating specific tapping points (Figure 14).

Figure 14. Mechanical adhesion

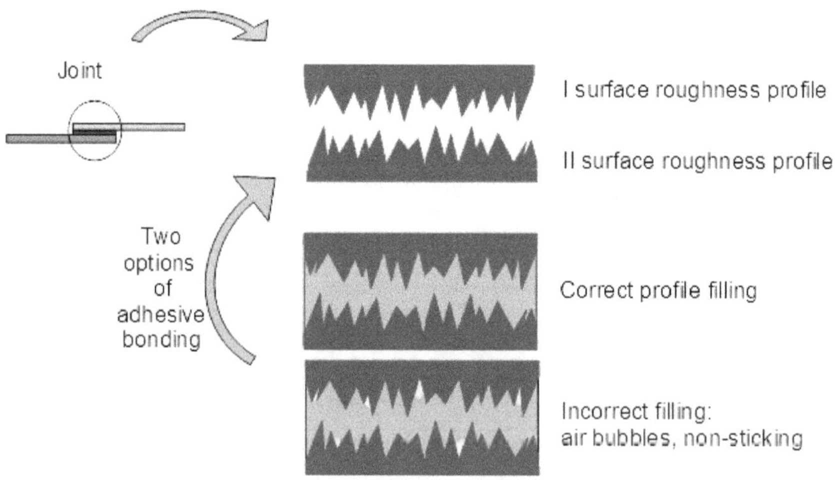

In the theoretical model it is assumed that pores and pits have irregular capillary shapes, and that their surface is not contaminated. The magnitude of pits plays a role in glue penetration, as glue viscosity limits its penetration. Roughness of real objects designated for gluing differs, however, from the ideal model. Properties of an adhesive and the material it penetrates need to be taken into consideration. Due to the fact that many production processes are done under time pressure, filling the pits with an adhesive may be incomplete. This, in turn, leads to production of elements with possible defects of adhesive bonds, as in the places designed for glue there remains air. Tapping points-bondages comprised of glue with rough material do not undergo decohesion on the boundary of phases under the influence of external loads. Decohesion, however, may occur inside the weakest material. In such a case, the bonding strength is determined by the strength of the adhesive or the bonded material (unlike the phenomena that take place on the boundary of 'solid-liquid' phases) (Żenkiewicz M., 2000).

Mechanical adhesion is, therefore, based on the assumption that despite the dependency of bonding strength of the glue or material, this strength can still be better. A decisive factor is a significantly greater contact area of the glue with the bonded, rough element, than in the case of smooth elements (Figure 15).

Figure 15. The influence of the roughness and thickness of the adhesive on the quality of the adhesive joint: a) the exclusive filling of cavities in the material with adhesive allows the adhesive to anchor in the cavities of the surface, but may not be sufficient, b) the anticipated greater thickness of the adhesive allows for the formation of adhesion and cohesion forces in the connection

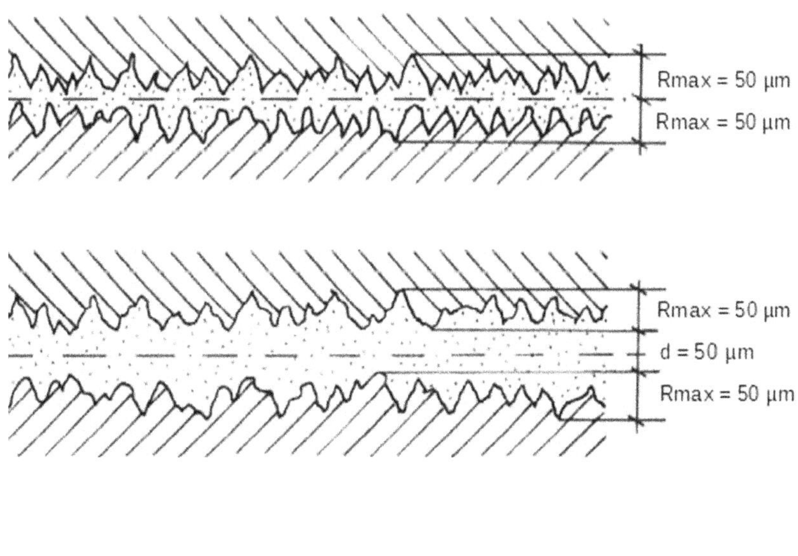

Additionally, if the forces that cause deformation occur, the form of these deformations is different as the pits and the layer of adhesive and the material are subjected to larger deformations (larger amount of energy is absorbed), than in the case of smooth parts.

Therefore, the final conclusion is that increasing the surface leads to the increase of the number of places for potential influence of an adhesive and, therefore, to the increase of strength properties (Żenkiewicz M., 2000).

CONCLUSION

The presented description of bonding shows that it is a complex and a multi-stage process. Each stage is dependent on many factors. The problem of choosing the right adhesive method concerns not only the preparatory operations, such as proper surface treatment and the preparation of the adhesive mass but also aims to ensure proper curing conditions for the joint and finishing the joint. The correct implementation of the individual stages of this bonding technique allows to achieve favorable adhesion conditions. According to the manufacturing needs and measurement possibilities, it can be assessed based on the energy state of the surface and its roughness - in both cases the control and measurement methods are quite simple. Ultimately, a specific condition of the surface and the conditions under which the bonding takes place can be verified, both in laboratories and industrial plants, by testing the strength of the bond and resistance to various conditions existing during the operation of the glued product.

REFERENCES

Brockmann, W., Geiß. P. L., Klingen, J., & Schröder, B. (2009). Adhesive bonding. Materials, applications and technology. Wiley-VCH Verlag GmbH&Co. KGaA.

Czaplicki, J., Ćwikliński, J., Godzimirski, J., & Konar, P. (1987). Klejenie tworzyw konstrukcyjnych [Bonding of construction materials]. Warszawa: Wydawnictwa Komunikacji i Łączności.

Deb, A., Malvade, I., Biswa, P., & Schroeder, J. (2007). An experimental and analitycal study of the mechanical behavior of adhesively bonded joints for variable extension rates and temperatures. *International Journal of Adhesion and Adhesives*, *28*(1-2), 1–15. doi:10.1016/j.ijadhadh.2007.02.004

Dutkiewicz, E. (1998). Fizykochemia powierzchni [Surface physicochemistry]. Warszawa: Wydawnictwa Naukowo-Techniczne.

Dziubak, C., & Rećko, W. (2009). Szok cieplny w tworzywach ceramicznych [Heat shock in ceramic materials]. *Prace – ISCMOiB, 4.*

Farnell. (2019) *Epoxy*. Retrieved from https://pl.farnell.com/araldite/araldite-2021-1-50ml/epoxy-adhesive-cartridge-yellow/dp/2676048

Fraunhofer. (2015). *European Adhesive Engineer.* Educational materials Fraunhofer IFAM.

Gaukler, J. C., Fehling, P., & Possart, W. (2009). *Thermo-oxidative and hydrothermal ageing of epoxy-dicyandiamide adhesive in bonded stainless steel joints*. 5th International EEIGM/AMASE/FORGEMAT Conference on Advanced Materials Research, IOP Conferences Series: Materials Science and Engineering 5. 10.1088/1757-899X/5/1/012002

Godzimirski, J. (2009). *Problemy klejenia konstrukcyjnego* [Problems of structural gluing]. Technologia i Automatyzacja Montażu, 1.

Godzimirski, J., Komorek, A., Rośkowicz, M., Smal, T., & Tkaczuk, S. (2010). Tworzywa adhezyjne. Zastosowanie w naprawach sprzętu technicznego [Adhesive materials. Application in technical equipment repairs]. Warszawa: Wydawnictwa Naukowo-Techniczne, Warszawa 2010.

Godzimirski J., Kozakiewicz J., Łunarski J., Zielecki W. (n.d.). *Konstrukcyjne połączenia klejowe elementów metalowych w budowie maszyn* [Structural adhesive joints of metal elements in mechanical engineering]. Rzeszów: Oficyna Wydawnicza Politechniki Rzeszowskiej.

Grant, L. D. R., Adams, R. D., & da Silva, L. (2009). Effect of the temperature on the strength of adhesively bonded single lap and T joints for the automotive industry. *International Journal of Adhesion and Adhesives*, *29*(5), 535–542. doi:10.1016/j.ijadhadh.2009.01.002

Hebda, M. (2007). Procesy tarcia, smarowania i zużywania maszyn [Friction, lubrication and wear processes of machines]. Warszawa: Wydawnictwo Instytutu Technologii Eksploatacji – PIB.

Karny, M. (2016). Połączenia klejone w strukturach kompozytowych – metodyka badan [Glued joints in composite structures - research methodology]. Warszawa: Prace Instytutu Lotnictwa.

Kłonica, M., & Kuczmaszewski, J. (2011). Ozonowanie jako metoda aktywowania warstwy wierzchniej materiałów konstrukcyjnych w operacjach klejenia [Ozone treatment as a method of activating the surface layer of construction materials in gluing operations] [Automatyzacja Montażu.]. *Technologia*, i.

Kuczmaszewski, J. (1990). *Technologia śmigłowców. Teoria i technika klejenia* [Helicopter technology. Theory and technique of gluing]. Lublin: Wydawnictwo Politechniki Lubelskiej.

Loctite. (2019). *Kleje strukturalne Katalog* [Structural adhesive Catalog]. Retrieved from https://www.albeco.com.pl/data/catalogueb/Loctite%20-%20Kleje%20Epoksydowe%20Loctite%20Hysol.pdf

Muller, M., & Valasek, P. (2010). *Adhesive bonds degradation*. Engineering For Rural Development.

Porębska, M., & Skorupa, A. (1997). Połączenia spójnościowe [Coherent connections]. Warszawa: Wydawnictwa Naukowe PWN.

Przygocki, W., & Włochowicz, A. (2011). Fizyka polimerów [Polymer physics]. Warszawa: Wydawnictwo Naukowe PWN.

Puff, T., & Sołtys, W. (1980). Podstawy technologii montażu maszyn i urządzeń [Fundamentals of machine and device assembly technology]. Warszawa: Wydawnictwa Naukowo-Techniczne.

Rudawska, A. (2016). Oprzyrządowanie w technologii klejenia [Tooling in gluing technology]. Lublin: Agencja TOP.

Rudawska A., Chruściel M. (n.d.). *Wpływ sposobu przygotowania powierzchni na wytrzymałość połączeń klejowych wybranego stopu aluminium wykorzystywanego w przemyśle lotniczym* [The impact of surface preparation on the strength of adhesive joints of a selected aluminum alloy used in the aviation industry]. Technologia i Automatyzacja Montażu, 2

Rudawska, A., & Dębski, H. (2011). Ocena wytrzymałości połączeń klejowych blach ze stopu aluminium z wykorzystaniem analizy numerycznej MES [Assessment of adhesive strength of aluminum alloy sheets using FEM numerical analysis]. *Eksploatacja i Niezawodność*, *1*, 4–10.

Rudawska, A., & Jacniacka, E. (2009). Analysis for determoning surface free energy uncertainty by the Owen-Wendt method. *International Journal of Adhesion and Adhesives*, *29*(4), 451–457. doi:10.1016/j.ijadhadh.2008.09.008

Rudawska, A., & Kuczmaszewski, J. (2005). *Klejenie blach ocynkowanych* [Bonding of galvanized sheets]. Lublin: Wydawnictwo Politechniki Lubelskiej.

Szabajkowicz, W., & Towt, A. (1994). Zastosowanie w montażu klejów i tworzyw klejopodobnych o właściwościach metalicznych [Application in the assembly of adhesives and adhesive-like materials with metallic properties]. *Technologia i Automatyzacja Montażu*, *2*, 6–9.

Wiącek, Z. (1987). Badania wpływu stanu powierzchni metali lekkich na jakość połączeń adhezyjnych [Research on the impact of light metal surface condition on the quality of adhesive joints]. Lublin: Wydawnictwo Politechniki Lubelskiej

Żenkiewicz, M. (2000). Adhezja i modyfikowanie warstwy wierzchniej tworzyw wielkocząsteczkowych [Adhesion and modification of the surface layer of high-molecular plastics]. Warszawa: Wydawnictwa Naukowo-Techniczne.

Zielecki, W. (2005). *Konstruowanie klejowych połączeń czopowych walcowych* [Construction of glue cylindrical spigot joints]. Technologia i Automatyzacja Montażu, 1.

Zielecki, W. (2004). *Destrukcja zmęczeniowa połączeń klejowych* [Fatigue destruction of adhesive joints]. Technologia i Automatyzacja Montażu No. 3-4.

Chapter 4
Technological Methods of Surface Pre-Treatment:
Traditional Approach

ABSTRACT

This chapter presents the traditional, widespread ways of implementing the first stage of the bonding operation (i.e., the preparation of materials surface), the importance of which has already been presented in Chapter 3. A number of different variants of this operation were described, depending on the bonded material, technical capabilities, and the skills of the manufacturer. The examples of particular fragments of surface preparation operations were given as well as the diversity of chemical substances, tools, equipment for work stands necessary for proper surface preparation. Other unconventional methods of surface preparation were briefly presented.

STAGES OF SURFACE PREPARATION PROCESS

The adhesive bonding process and its basic requirements were already presented in chapter 3.

A decisive condition for obtaining a strong bond is good surface pre-treatment before adhesive bonding. It is an initial stage of the technological process of adhesive bonding and it must ensure (da Silva L., Lopes M., 2009):

DOI: 10.4018/978-1-7998-4634-5.ch004

- removal of all surface contamination in the form of dirt, dust, loosely joined corrosion layers and greases,
- obtaining proper development of surface,
- obtaining good surface activation.

This stage is preceded by the analysis of the type and properties of materials designated for adhesive bonding. It is them that determine the choice of a method, means and technological parameters of adhesive bonding (Rudawska A., Chruściel M., 2011)

The most important stage of a technological adhesive bonding process requires that a series of activities is performed before applying the resin compound (Rudawska A., Kuczmaszewski J., 2005):

1. degreasing
2. cleaning in acid or alkaline bath
3. rinsing in water
4. drying
5. surface special processing – mechanical, chemical, electrochemical treatment, plasma digestion, corona treatment, ozoning, laser treatment
6. rinsing
7. drying (+ application of primer)

A specific method of preparing the surface influences the strength of the bond obtained after adhesive bonding of the components. The choice of a particular method depends on several factors (Rudawska A., 2010; Ebnesajjad S., 2008):

- prospective operation conditions of the glued part,
- limitations concerning the applied material,
- limitations resulting from the shape and structure of the object,
- manufacturing conditions – the available tools, equipment, processing and measuring devices, lead times, etc.

Degreasing

Metal products are usually characterized by high smoothness and cohesive surface structure, but they are very frequently covered with greases that comprise a trace after processes, for example, stamping, rolling, forging or pressing. Sometimes they also have a layer that protects them from corrosion; hence it is important to remove grease contamination. Most frequently this process is performed with trichloroethylene or tetrachloroethylene in the form of vapor. Metal components of small sizes are placed in chambers of special washers, where vapor of the aforementioned solvents rinse them in a continuous manner, by condensing on their surface (Rudawska A., 2014)

The process of cleaning in solvent vapors lasts until the temperatures of vapors and of the components being cleaned equalize. If the grease layer is significantly bigger and it is necessary to lengthen the cleaning time, the elements are cooled down by removing them from the washer chamber or they are hung in a chamber equipped with special tubular hangers cooled down in a continuous manner with a flow of water stream directed towards the inside of the pipes. If the components being cleaned from grease are big and do not fit inside washer chambers their surfaces are rubbed with a cloth soaked with a solvent. Such cloths must be clean and must not include soap traces; such cloths are also frequently replaced. Also in workshop conditions it is possible do degrease components by means of rubbing with a cloth. Petrol, acetone, benzene or ethanol is frequently used instead of trichloroethylene or tetrachloroethylene. These solvents, however, are less effective than tetrachloroethylene (tetrachloroethylene is also less toxic than trichloroethylene) (Thrall W., Shannon W., 1985).

Sometimes the surfaces of components designated for adhesive bonding are so contaminated with burnt grease that it is recommended to initially soak them in solvents and then wash by using tools and mechanical devices. In such cases it might also be effective to use an ultrasonic or electrolytic washer (Brockmann W., 2009).

Cleaning in Acid or Alkaline Bath

Depending on the type of cleaned metal, acid or alkaline baths are applied. It is recommended that steels and titanium alloys that have oxide layers or scales should be etched in acid baths; aluminum alloys are etched, by principle, in significantly cheaper alkaline baths containing detergents which effectively remove organic contamination. Information on the ingredients of many alkaline cleaning agents is not provided by manufacturers; in such a case it is recommended to follow their recommendations or determine the ingredients of such a bath in an experimental way (Thrall W., Shannon W., 1985).

Rinsing in Water

A rinsing operation may be performed in various ways. Sometimes it is recommended to use cold water, in other cases – the hot one. Some authors accept using municipal water, while others – demineralized or even distilled water.

One may also encounter different recommendations concerning the rinsing technique – with the spray method (firstly with cold water, then with hot water) or the immersion method.

In some cases attention is also paid to possible contamination with organic content. The content of metallic ions in water is considered as beneficial in the case of aluminum alloy rinsing. In such a case unfavorable oxides weakly joined with the surface are not created. They can weaken the prospective adhesive bond, making it easy for an adhesive to get detached together with oxides. In the case of aluminum alloys, spray rinse with cold water is recommended and it should be performed effectively (in short time). Such rinsing is aimed at removing the traces of etching or cleaning baths because of the fact that they may be a source of corrosion or cause changes in adhesive chemical properties (Rudawska A., Kuczmaszewski J., 2005).

Drying

An adhesive is applied on the surface that is clean and dry. After taking them out of bath, the components are air dried in ambient temperature in a clean air environment, or mechanical ventilation or chamber driers are used (Kuczmaszewski J. 2011). In most cases drying should not take place in temperature exceeding 55°C (Czaplicki J. & Others, 1987).

MECHANICAL METHODS OF SURFACE PRE-TREATMENT AND SURFACE CONDITIONS

Surface pre-treatment is one of the first and most important technological stages in the adhesive bonding process. It is preceded by the analysis of properties, type and geometrical structure of material surface for adhesive bonding, as the choice of an appropriate surface pre-treatment method depends on these (Rudawska A., 2014). Surface pre-treated for adhesive bonding should be characterized by cleanliness, good surface energy and it should be developed in a particular way.

Strength properties of bonds depend on proper pre-treatment of the surface designated for bonding. In manufacturing conditions obtaining the aforementioned results may be achieved, among others, with mechanical methods.

Methods of mechanical pre-treatment of surface are:

- **manual:** done with using, for example, abrasive paper of various granulation. Abrasive paper may be attached to a wooden or plastic block (cube) so as to make holding easier. A cuboidal block ensures that the pressure exerted by the paper on the surface is uniform and equal. Another method is to make use of grinding paste and the so-called roughing by using brushes and mechanized tools, such as flap disc grinders (Rudawska A., 2017).
- **mechanical:** made with machine tools – grinding, polishing, sand-blasting, milling etc. In reality components are very often sand-blasted by using blast media, such as fused alumina, ceramic or mineral particles, thrown from the device nozzle under particular pressure. Particle sizes may vary in size, e.g., of fused alumina – 180-325 μm, silicon carbide or quartz sand – 250-310 μm. Blasting takes place in special chambers, where a pre-treated object is placed and the nozzle throws a stream of particles and directs it in the direction of the machined surface. There are devices with nozzles operated manually, but there are also ones that are equipped with numerous nozzles in many places, without the necessity to move them. The first solution is suitable for small plants, where the production range is narrow and manual machining is sufficient. It is required, however, from the employees to have qualifications so as to ensure even nozzle movement (in particular the travelling speed regarding the parts, the distance from the nozzle outlet to the surface, number of machining repetitions, control of parts' consistency and dimensions of parts). The second presented solution is better in the case of large plants, in which various elements of varied shapes are manufactured. Full mechanization provides the possibility of fast machining, repeatability of results arising from programming possibilities and machining cycle replication. In such a case sand-blasting chambers have frequently bigger sizes.

Because of the internal stresses generated in all sand blasting processes it is essential to select appropriate machining parameters, such as grain size, nozzle air pressure and machining time, with regard to the material type, its hardness, shape, structure and part stiffness (thin-walled elements are particularly susceptible). One type of mechanical processing with the use of loose blasting agent is water-jet blasting (Brockmann W., 2009).

Slightly worse surface adhesive properties are obtained after grinding with grinding wheels of small grains. Additionally, before such machining the surface needs to be degreased, it must not have any notches (metallic inclusions, e.g., of higher hardness), technological notches (remaining ridges, burrs) so as not to damage the grinding wheel.

Another machining method is brushing, after which surface condition is worse than after grinding. Nevertheless, brushes are used instead of grinding wheels. Brushes are elastic (hence deterioration of surface properties); the surface, like before grinding, should be degreased. Machining result depends on brush speed, its size and angle in relation to the surface. Because of possible component deformations, brushing is not recommended for floppy components with big but thin parts. The very process of machining may be very noisy.

Machining with abrasive tools reinforces adhesive properties of surface layers, mainly as a result of surface development, i.e., increasing the area of the true adhesive contact, as well as removal of organic and inorganic contamination from the surface. For this reason grinding, milling, brushing, sand-blasting and other processes are used (Domińczuk J., 2001) (Zielecki W., 2007).

When selecting an abrasive tool for machining one should pay special attention to grain size. Tools with a bigger grain size ensure better removal of excessive material, including the physisorption layer.

Sometimes the resultant strength of bonds is associated only with the technological parameters of the process, such as the size of fused alumina grains, granularity of the abrasive paper, etc. (Zielecki W., 2010; Łunarski J., Zielecki W., 1994). Many researchers, however, also include surface roughness parameters in accordance with the well-known theory of mechanical adhesion. Measurement of surface roughness is relatively easy to perform in manufacturing conditions. One should not forget, however, that technological guidelines should refer to a particular material type.

According to the aforementioned mechanical theory of adhesion, height parameters for indentations and elevations are vital with respect to adhesive penetration (Łunarski J., Zielecki W., 1994). Theoretically, for adhesive bonding the surface of Rm (maximal roughness height) in the range from 7 to 25 μm for A283 Gr.C steel is appropriate. However, (Uehara k., Sakurai M., 2002), for example, use a very popular parameter of Ra – average profile arithmetic deviation- 1% for carbon steel, providing only optimal Ra conditions in the range of μm.

The results of experimental works indicate, however, that these are not sufficient parameters. For example, in the work of (Zielecki W., 2007; Łunarski J., Zielecki W., 1994) surfaces were additionally characterized and Rz (roughness height of 10 points) was used. It was stated that by means of abrasive machining methods, such as polishing, grinding and brushing, the surface of one-directional structure is formed. After machining with diamond grit wheel the structure is multi-directional, while in the case of abrasive blasting (sand-blasting) it is most beneficial, non-directional and pointwise. During machining there occurs removal of oxide layers from the surface, for example, of the rolled metal sheets. An exception to that is machining with an abrasive disc or shot peening, where the machining process is random and there may appear areas with an oxide layer on the surface that were not removed. The result of machining by polishing, grinding and brushing is the surface the Rz of which does not exceed 15 μm. On the other hand after abrasive blasting or machining with an abrasive disc the Rz may vary in the range of 20-35 μm. The level of surface development was characterized by the lr parameter – the profile ratio that defines the relative elongation of a real profile in relation to the elementary distance in surface roughness measurements.

Surfaces after abrasive machining and shot-peening may have lr not higher than 1.03, while after shot blasting 1.04-1.07 (Łunarski J., Zielecki W., 1994). It was stated that surfaces of high roughness do not have to maintain significant profile development. The reason for low effectiveness, for example, of abrasive machining, may be high temperature of machining, which in result initiates the process of reforming the oxide layer on the surface. Another phenomenon that occurs during blasting may be the time of processing. It has a significant meaning in the case of thin-wall components, when too long exposure to the influence of blasting medium may cause occurrence of detrimental stresses and even material deformation. Then, as a result of shortening the processing time, surface cleaning can be too short, leading to insufficient surface development. In the work of (Zielecki W. & Others, 2011), on the other hand, it is stated that an additional, advantageous parameter for surface development description are spatial parameters (hybrid ones), as well as arithmetic average curvature of Ssc vortices, developed interfacial area ratio Sdr, root mean square gradient Sdq. In order to determine surface development by means of these parameters, however, measurement devices that enable measurement of the structure in spatial arrangement are necessary (they are not always available in workshop conditions).

CHEMICAL AND ELECTROCHEMICAL METHODS OF SURFACE PRE-TREATMENT

Apart from mechanical methods of surface processing, there are other options of processing the surface designated for adhesive bonding. Chemical and electrochemical methods constitute another numerous group of possible surface pre-treatment methods.

Many materials, such as aluminum and aluminum alloys, cannot be subjected to abrasive machining. It results from their hardness – machining is ineffective due to quick wear of tools and chemical phenomena taking place on metal surface. A disadvantageous phenomenon is corrosion; some alloys are susceptible to corrosion and cannot be exposed to contact with water, high temperature or other ambient factors. In such a case chemical and electrochemical methods are suitable, as they make it possible to form a beneficial layer of oxides on metal surface, of low thickness and strongly connected to the basis and with good adhesive properties (David E., 2004).

Chemical and electrochemical methods are useful for pre-treatment of elements made of aluminum, magnesium and titanium alloys.

Three basic methods of chemical processing are distinguished:

- **acid bath etching (ABE):** the process of surface pre-treatment that relies on removing natural oxides from metal surface,
- **anodizing in sulphuric acid (ASA):** it is a multistage process: it starts with initial etching in acid bath, then electrochemical anode oxidizing and sealing are performed. A layer of oxides is obtained; it ensures high durability of adhesive bonds and maximum protection against corrosion,
- **anodizing in chromium acid (ACA):**it is initial etching in acid or alkaline bath and then electrochemical anode oxidizing.

  Acid etching is applied for:
  - surface preparation of standard samples for assessing the strength of adhesives for metals,

- surfaces that do not require anticorrosion protections or layers with coat paints applied after adhesive bonding,
- surface pre-treatment before anode oxidizing in sulphuric or chromium acid.

Anodizing in sulphuric acid is used for:

- pre-treatment of component surfaces, for which protection with coat paint after adhesive bonding is insufficient,
- surface pre-treatment of components of thickness over 0.25 mm.

Anodizing in chromium acid is used for:

- surface pre-treatment of components that require protection against corrosion, for which anodizing in sulphuric acid is not recommended due to shapes that hinder electrolyte removal,
- surface pre-treatment of components that require thin anode layers, high fatigue strength and components of the thickness below 0.25 mm.

The procedure for selection of a proper method is presented below:
In the presented procedure the following actions are distinguished:

1. Choice of the method of chemical pre-treatment – ABE, ASA or ACA.
2. Degreasing – for all methods – with solvents, emulsion degreasing or in solvent vapors; the reaction should be neutral (app. pH 7). Alkaline reaction (above pH8) is not recommended due to surface over-etching. The solvent should not be contaminated with oil or grease.
3. Acid etching (specific for ABE, or pre-treatment stage for AS and TC) – different solutions are applied for acid etching:
   Solution 1: H_2SO_4 (150 cm^3) + CrO_3 (50 g) + water (850 cm^3)
   Solution 2: H_2SO_4 (150 cm^3) + $Na_2Cr_2O_7$+H_2O (75 g) + water (850 cm^3)
   The bath is heated to the temperature of app. 60°C and parts are immersed for about 20 min. If acid etching was selected then the next stage is stage 8.
4. Alkaline etching – it is allowed instead of acid etching before anode oxidizing in chromium acid.
   The bath takes place in NaOH of 2% in the temperature of 75°C in the time of 5 min.
5. Clearing – after alkaline etching in water bath solution of HNO_3 (technical) 20%, in the temperature of app. 20 °C and the time of 2 min.
6. Anode oxidizing takes place in
   - in the case of ASA:

water solution with H_2SO_4 (cz.15 g) of 1%, in the temperature of 10-15°C, anode current density of 1.5±0.2 A/dm^3 in the time of app. 20 min.; the amount of the passed electricity: 30-33 A min/dm^2, direct current voltage on bath rails of 20 ±2 V,

- in the case of selecting ACA:

  water solution with CrO_3 (tech.) 5-10%, in the temperature of app. 40°C, in the time of 20 min.; direct current voltage within first 10 minutes changes smoothly by 5V, from 0 to 40V, then it maintains 40V for 20 minutes, then it increases to 50V within 5 minutes and maintains 50V for last 5 min.

7. Sealing – for anode sealing in sulphuric acid – dipping a part in the bath: Water + $K_2Cr_2O_7$ (50-100 g/dm^3), bath pH 6.0±0.2, temp. 90±5°C, time 6±1 min.
8. Stirring – in all methods. Stirring should be performed during the stage of clearing (5), by applying compressed air without oil traces. During the stage of degreasing, acid etching and anode oxidizing (2), (3) and (6) the solutions are stirred according to needs, until obtaining the required temperature.
9. Bath surface cleaning – each time after stating the presence of contamination on the solution surface in the bath.
10. Component cleaning – in running municipal water after the activities of (2), (3), (4), (5) and (6).
11. Drying – in the time of at least 3 hours in ambient temperature or 15 minutes in the temperature of 105±5°C. For all solutions distilled water is used (Czaplicki J. & Others, 1987; Habenicht G., 2009; Thrall W., Shannon W., 1985).

Other types of chemical processing are used in manufacturing. Sometimes production plants develop their own procedures for the needs of particular products. Aviation plants and machining of wing parts by alodine treatment may serve as an example.

Alodine treatment is a fast, non-current process of applying a convertible chromate layer. It improves corrosion resistance of aluminum and its alloys; it is also applied before painting and adhesive bonding operations. The layer may be applied manually or by immersing the component in a bath. The process relies on applying the so-called alodine (water solution of powder, e.g., 1200S of the Henkel Surface Technologies Company). After the process of alodine treatment the chemical layer obtains the thickness of up to 1 μm, it is brittle and susceptible to mechanical damage. The alodine treatment process takes place in the temperature of 18-24°C, by moistening from 1 to 5 min. (manually) or immersion from 15 s to 3 min. The layer of oxides formed this way has nice, yellow or greenish color (Sękowski S., 1987).

In laboratory practice, for the purpose of comparative studies sometimes other simplified methods of surface chemical pre-treatment are used.

Example

Pre-treatment of aluminum surface in experimental conditions according to (Rudawska A., Chruściel M., 2011):

Variant 1:

Degreasing – Loctite 7061 (substance that contains acetone) – after several applications it degreases, then wipe with a clean cloth, the last layer is to remain without wiping for free evaporation.

Variant 2:

Etching in water solution of NaOH of the concentration of 40-60g/l, dipping elements in a bath of the temperature of 40-60° for 3-4 min., taking out and rinsing in hot, running water. Leaving for free evaporation.

Variant 3:

Chromating – in the solution of acids of $Na_2Cr_2O_7$ in the amount of 5.5g/l; Na_2SO_4 in the amount of 4 g/l; HNO_3 in the amount of 1.5 (pH 1.4-1.6) for about 30 s.

Variant 4:

Anodizing – electrolyte of 20% H_2SO_4, parts dipped in a bath of the temperature of 40°C and voltage of 50V for 35-40 min.

OTHER METHODS OF SURFACE PRE-TREATMENT BEFORE ADHESIVE BONDING

Apart from mechanical and chemical surface pre-treatment, other methods are used slightly less often. These may include, for example:

- **Ozonization**: due to strong oxidizing and radical-generating behavior, it is used in plastics industry, construction, aviation, metallurgy and ceramic industry. It is considered an ecological method due to lack of industrial and toxic waste, as well as an economical one due to low costs of creating ozone and operation of equipment; objects are placed in a chamber with ozone (of particular concentration) and kept for particular time (Kłonica M., Kuczmaszewski J., 2011; Kwiatkowski M. & Others, 2009),

- **Ion Bombardment**: it is used to modify surface parameters of polymers and it is beneficial owing to the fact that it reacts with all material components, proportionally to their electron share. The method poses some problems, as the secondary effect of radiation are not only cross-linking, but also degradation. Ion bombardment may, however, be used in a clever way thanks to new properties of the polymer, such as improvement of wettability and adhesion, thermal and chemical stability, resistance to abrasive wear and other (Bieliński D.M. & Others, 2013),
- **Low-Pressure Plasma Interaction**: operation conducted in vacuum (partly), reacted gases, e.g., oxygen, hydrogen, fluorine, are needed. Gases are transferred into an energy-rich state plasma, by microwave stimulation with the possibility of chemical surface modification. Gases are fed into the plasma chamber with the adherends to be pre-treated (Habenicht G., 2009). The pressure in the chamber during the gas flow is stable on the level of app. 40 Pa, the operation time is app. 60 min. (Leahy W. & Others, 2001),
- **Corona Discharge Treatment:** the technique is similar in principle to low pressure plasma but the plasma is generated in air at atmospheric pressure. The corona is generated by applying high voltage (up to 30 kV) at the frequencies ranging from 9-50 kHz to an electrode separated from an earthed table by an air gap (Bieniak D., 2009),
- **Flame Treatment:** treatment that requires the following gases: propane/butane or acetylene with excess oxygen. Used to chemical and physical surface modification with oxidative effects. The material is gently touched by an open flame. This induces an oxidative effect that improves the adhesion properties, especially in the case of nonpolar plastics. The flame treatment time is in the range of seconds, the distance of the flame to surface should be approximately 5-10 cm. In the case of the thermoplastics, care should be taken that surface melting is avoided. Flame treatment does not require a special chamber, and can be conducted manually. This method is also safe for the natural environment (Habenicht G., 2009).

CONCLUSION

From the presented review of the technological possibilities of performing the first stage of the bonding operation, which is the preparation of the material surface, it follows that this stage is extremely diverse, giving a variety of options for the choice of an appropriate machining technique. This has several advantages, as the area of bonding applications increases due to materials and manufacturing capabilities. However, it should be particularly emphasized that an undoubted disadvantage of traditional methods is the need to use different types of chemicals, their mixtures, clean water, use of many different tools and technical equipment. It also must be remembered that after the completion of technological activities, solid waste and contaminated liquids remain in the production process, which often poses a threat to human health and impose an environmental burden, and their amount may be significant.

REFERENCES

Bieliński, D. M., Lipiński, P., Okrój, W., & Klimek, L. (2013). Selected examples of the modification of the Surface layer of polymers by ion bombardment. *Inżynieria Materiałowa*, *27*(6), 1337–1342.

Bieniak, D. (2009). Preparing polymer substrates for bonding. Reiforced Plastics. Elsevier Ltd.

Brockmann, W., Gei, ß. P. L., Klingen, J., & Schröder, B. (2009). Adhesive bonding. Materials, applications and technology. Weinheim: Wiley-VCH Verlag GmbH&Co. KGaA.

Czaplicki, J., Ćwikliński, J., Godzimirski, J., & Konar, P. (1987). Klejenie tworzyw konstrukcyjnych [Bonding of construction materials]. Warszawa: Wydawnictwa Komunikacji i Łączności.

da Silva, L., & Lopes, M. (2009). Joint strength optimization by the mixed-adhesive technique. *International Journal of Adhesion and Adhesives*, *29*(5), 509–514. doi:10.1016/j.ijadhadh.2008.09.009

David, E., Lazar, B., & Armeanu, A. (2004). Surface modification of polytetrafluoroethylene for adhesive bonding. *Journal of Materials Processing Technology*, *157-158*, 284–289. doi:10.1016/j.jmatprotec.2004.09.045

Domińczuk, J. (2001). *Steel adhesive properties changes after the chosen ways surface layer preparations*. Kraków: Postępy w Technice Wytwarzania Maszyn.

Ebnesajjad, S. (2008). *Adehesives Technology Handbook*. Norwich: Wiliam Adrew.

Habenicht, G. (2009). *Applied adhesive bonding. A practical Guide for Flawless Results*. Weinheim: Wiley-VCH Verlag GmbH&Co. KGaA.

Kłonica, M., & Kuczmaszewski, J. (2011). *Ozonowanie jako metoda aktywowania warstwy wierzchniej materiałów konstrukcyjnych w operacjach klejenia* [Ozone treatment as a method of activating the surface layer of construction materials in gluing operations]. Technologia i Automatyzacja Montażu, 2.

Kuczmaszewski, J. (2011). *Efektywność wytwarzania elementów lotniczych ze stopów aluminium i magnezu* [The efficiency of manufacturing aviation components from aluminum and magnesium alloys]. Komputrowo zintegrowane zarządzanie.

Kwiatkowski, M., Kłonica, M., Kuczmaszewski, J., & Ozonek, J. (2009) The use of the ozonation proces to modify energetic properties PA6 polyamide surface layer. Lublin: Polska Akademia Nauk, Monografie No. 59, T.2.

Leahy, W., Barron, V., Buggy, M., Young, T., Mas, A., Schue, F., ... Bridge, M. (2001). Plasma Surface Treatment of Aerospace Materials for Enhanced Adhesive Bonding. *The Journal of Adhesion*, *77*(3), 215–249. doi:10.1080/00218460108030739

Łunarski, J., & Zielecki, W. (1994). Wpływ struktury geometryczne powierzchni na wytrzymałość na ścinanie połączeń klejonych [Impact of geometric surface structure on shear strength of glued joints]. *Technologia i Automatyzacja Montażu*, *2*, 13–16.

Rudawska, A. (2010). *Wpływ sposobu przygotowania powierzchni na wytrzymałość połączeń klejowych blach ze stali odpornej na korozję* [The impact of surface preparation on the strength of adhesive joints of corrosion-resistant steel sheets]. *Technologia i Automatyzacja Montażu*, *3*, 36–39.

Rudawska, A. (2014). Selected aspects of the effect of mechanical treatment on surface roughness and adhesive joint strength of steel sheets. *International Journal of Adhesion and Adhesives*, *50*, 235–243. doi:10.1016/j.ijadhadh.2014.01.032

Rudawska, A. (2017). Przygotowanie powierzchni do klejenia wybranych materiałów konstrukcyjnych [Preparation of the surface for gluing selected construction materials]. Lublin: Monografie-Politechnika Lubelska.

Rudawska, A., & Chruściel, M. (2011). *Wpływ sposobu przygotowania powierzchni na wytrzymałość połączeń klejowych wybranego stopu aluminium wykorzystywanego w przemyśle lotniczym* [The impact of surface preparation on the strength of adhesive joints of a selected aluminum alloy used in the aviation industry]. Technologia i Automatyzacja Montażu, 2.

Rudawska, A., & Kuczmaszewski, J. (2005). Klejenie blach ocynkowanych [Bonding of galvanized sheets]. Lublin: Wydawnictwo Politechniki Lubelskiej, 2005

Sękowski, S. (1987). Galwanotechnika domowa [Home electroplating]. Warszawa: Wydawnictwa Naukowo-Techniczne.

Thrall, W., & Shannon, W. (1985). *Adhesive bonding of aluminum alloys*. New York: Marcel Dekker, Inc.

Uehara, K., & Sakurai, M. (2002). Bonding strength of adhesives and surfaces roughness of joined parts. *Journal of Materials Processing Technology, 127*, 178-181.

Zielecki, W. (2007). *Wpływ rozwinięcia struktury powierzchni na wytrzymałość zakładkowych połączeń klejowych* [Impact of surface structure development on the strength of overlapping glue joints]. Technologia i Automatyzacja Montażu, 2-3.

Zielecki, W., Pawlus, P., Perłowski, R., & Dzierwa, A. (2011). *Analiza wpływu struktury geometrycznej powierzchni w układzie 3D na wytrzymałość połączeń klejowych* [Analysis of the influence of the geometric structure of the surface in a 3D system on the strength of adhesive joints]. Technologia i Automatyzacja Montażu, 1.

Zielecki, W., & Zielecki, K. (2010). Analiza MES wpływu struktury geometrycznej powierzchni na stan naprężeń w spoinie klejowej [FEM analysis of the influence of the surface geometrical structure on the stress state in the adhesive joint]. *Technologia i Automatyzacja Montażu, 4*, 49–51.

Chapter 5
Lasers:
Selected Aspects of Application in Production

ABSTRACT

This chapter describes basic issues concerning lasers (i.e., the phenomenon of laser action, features of laser radiation, construction of laser devices, and various types of lasers). Additionally, selected areas of laser applications in the modern industry were presented. Laser techniques related to the melting of the material, laser melting, various types of laser heat treatment, laser welding, drilling, cutting, and marking of materials with laser were discussed. In addition, laser strengthening techniques developed intensively in recent years and various additive manufacturing techniques using lasers were presented. The advantages and disadvantages of laser processing were also specified in this chapter.

PHENOMENON OF LASER ACTION

Atomic systems are in specific quantum states, so called energy levels. The change of these states can only take place in the form of a discontinuous, stepwise transition. The transition of a quantum system from one energy level to another is accompanied by absorption (absorption) or separation (emission) of a specific portion of energy. This portion is called quantum.

DOI: 10.4018/978-1-7998-4634-5.ch005

The quantum is equal to the difference in transition levels. The transition can be (Oczoś, 1988):

- radiant, where the energy absorbed or emitted by the quantum system is in the form of electromagnetic radiation,
- non-radiant, when energy is transferred to another atomic system (e.g., by collision of atoms or molecules in gases or solids).

Transitions in the extracted atom are exclusively radiant and occur according to the relation (quantum dependence of Bohr):

$$h\nu = E_2 - E_1$$

where:

$h\nu$ - radiation quantum (photon),
h - Planck constant, $h = 6.62517 \cdot 10^{-34}$, Js
ν - frequency of radiation
$E_2 - E_1$ - difference of energy levels between which the energy portion was passed through

The transition of the system from a lower energy level E_1 to a higher energy level E_2 takes place after supplying from the outside the quanta of radiation $h\nu$. The radiation is absorbed and the so-called absorption transition takes place.

The reverse phenomenon is emission. At the transition from a higher level of E_2 to a lower level of E_1, the quantum system gives the excess energy to the outside in the form of the quantum radiation $h\nu$. This is the so-called emission passageway.

If the energy level E_1 is the lowest possible, it is called the basic energy level (state). Then the state E_2 is the induced level (state) (Oczoś, 1988).

The natural tendency of the electron is to be at a lower energy level. If the induced electron is at a higher energy level than the basic level, it spontaneously passes to the basic level and emits a photon with energy $h\nu$. This is a spontaneous emission phenomenon. Such radiation is inconsistent (incoherent). The radiation of all common light sources is the result of spontaneous emission (Oczoś, 1988).

By interacting with matter (the so-called medium) with electromagnetic radiation of frequency:

Figure 1. The phenomenon of absorption (a) and emission (b) of energy in the atom

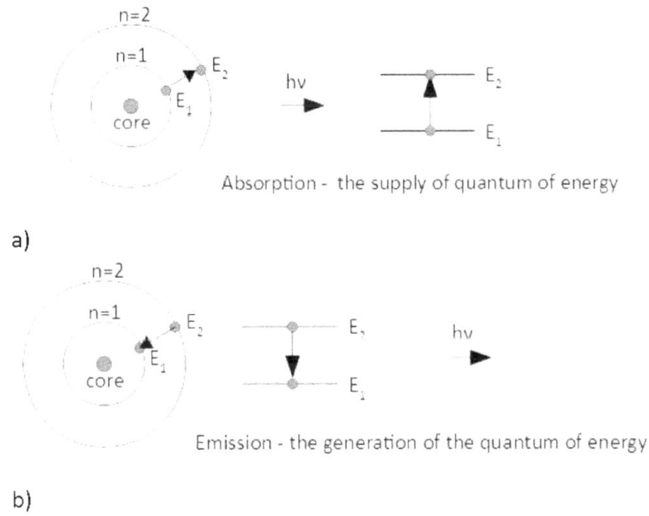

$\nu = (E_2 - E_1) / h$

two processes can be triggered:

- if the photon interacts with a particle which is at a lower energy (basic) level, then the radiation absorption takes place and the particle is forced to move to a higher energy level, or,
- if the photon interacts with a particle which is at a higher energy level, then this particle is forced to go to a lower energy level and emits a new photon of the same frequency as the photon originally interacting (falling on this particle). This phenomenon is called **forced emission** (Figure 2). Forced emission takes place in the same direction as the photon interacting with the particle. During forced emission, the frequency, phase and direction of radiation propagation are compatible. This radiation is called coherent.

Forced emission is the basic phenomenon on which laser operation is based In order for the so-called laser action to occur, it is necessary to bring the quantum system to the state in which the excited particles dominate. The emission of radiation is then possible (Oczoś, 1988).

Figure 2. Forced emission phenomenon

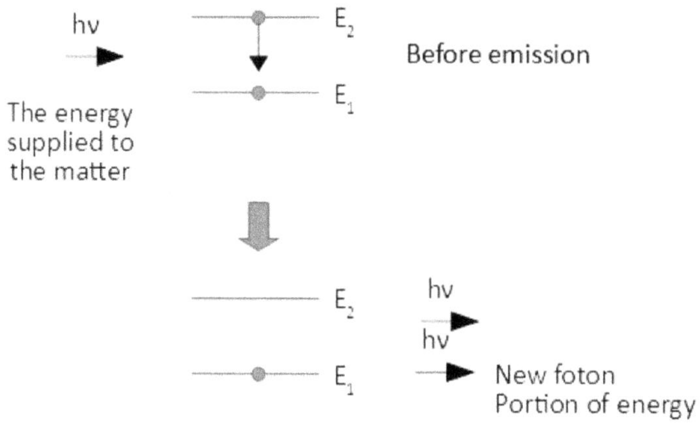

Population of Levels

In the state of thermal balance, lower energy levels are filled more. This condition is reversed. The process is called population inversion, and its consequence is a surplus of emitted photons over absorbed photons. The inversion of charge-carrier population is done by supplying energy to the system. In lasers, this is done by using the so-called pumping system.

The second condition for the occurrence of laser action, next to the inversion of fillings, is to amplify the radiation. The power of forced radiation must be greater than the loss of radiation that can occur due to diffraction, dispersion or reflection. Lasers use so-called optical resonators - resonance chambers in which an active medium is placed. With two mirrors perpendicular to the axis of the chamber, the radiation repeatedly passes through the active medium, thus triggering avalanche photon transitions. Laser action is triggered by axial, parallel or slightly deflected beams. Other beams have no possibility of adequate reinforcement.

All lasers are composed of an active medium and a pumping system (Ziętek, 2009).

The active medium is a matter whose particles will be induced; most often it is placed in the so-called optical resonator. It is to ensure that the power of the forced radiation was greater than the radiation losses. Such losses may occur due to diffraction, dispersion or reflection of radiation in the system. The resonator is usually made up of two mirrors perpendicular to the resonator

Figure 3. The condition of the proper operation of the resonator - wave nodes are formed on mirror surfaces

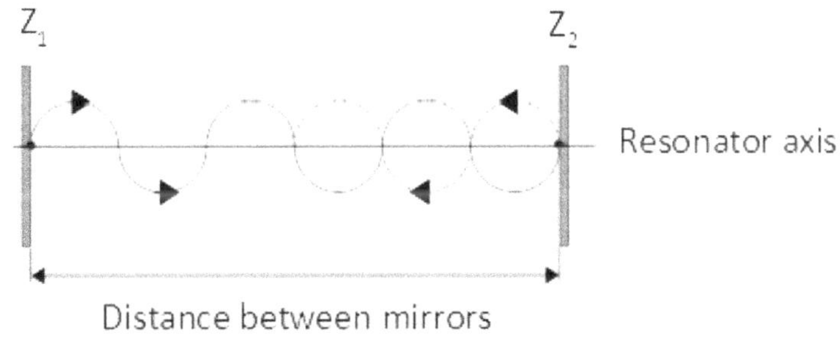

axis. For a resonator to work properly, the distance between mirrors must be equal to the total number of halves of the wave (Oczoś, 1988).

$L = q_n (\lambda_n/2)$

Meeting this condition ensures that the wave nodes are on the surface of the resonator mirrors. The optical resonator plays an important role in both the efficiency of the laser action and the parameters of the emitted beam.

Between mirrors arranged parallel to each other, a standing wave is formed due to interference of flat light waves spreading along the axis of the resonator in opposite directions (due to reflection from mirrors). This results in the so-called longitudinal vibration (mode). There are many types of longitudinal vibrations (modes). They do, however, fulfill the condition:

$L = q_1 (\lambda_1/2); L = q_2 (\lambda_2/2); ...; L = q_n (\lambda_n/2)$

The frequency range or wavelength range, which meet the above condition, creates the so-called wave spectrum (or frequency spectrum) resonance active medium, or the spectrum of laser radiation. The width of the spectral line of laser radiation is expressed in nm or Hz.

However, diffraction effects may occur at the ends of the mirrors. Then the amplitude and phases of the waves change. The so-called transverse vibrations (modes) are created. They characterize the distribution of electromagnetic field in the cross-section of the resonator. Transverse modes are marked with the symbol TEM_{mn} - Transverse Electro-Magnetic. The numbers m and n are integers, taking the values 0 or 1, and denote the row of transverse vibrations.

In the case of cavity machining, single mode TEM_{00} is used, in basic mode. The best conditions for focusing the beam are in this mode. The distribution of radiation intensity in the focused TEM_{00} beam is a normal Gauss distribution.

Figure 4. Gauss distribution of He-Ne laser intensity in a plane perpendicular to the propagation direction: a) 2D view, b) 3D view - single mode beam, c) 2D view, d) 3D view - multi mode beam

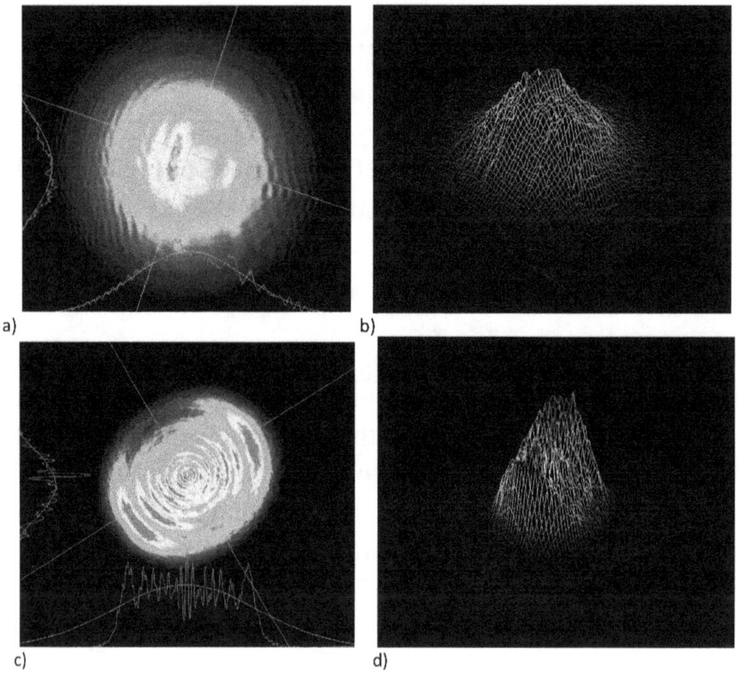

A pumping system is a device that supplies the energy needed to induce a laser's active medium. Pumping, i.e., forcing the inversion of planting levels by electrons can be done in different ways. One can distinguish (Ziętek, 2009):

- **Optical Pumping:** induction of a higher energy state is done selectively, by absorption of light emitted by another laser, light emitting diode or flashing lamp,
- **Chemical Pumping:** due to an exothermic reaction which causes selective inducing of the reaction products to the appropriate laser states,

- **Pumping By Non-Elastic Collisions With Electrons:** using ionizing radiation and accelerating electrons in an external electric field. Accelerated electrons collide with the atoms of an active medium and transfer their energy to them,
- **Pumping Through Flexible Collisions With Electrons:** these collisions lead to resonant excitation of atoms,
- **Thermal Pumping**: by providing the active center with thermal energy sufficient to induce it, pumping by injecting media into semiconductor lasers.

Figure 5. Example of generator (left) and fiber laser head (right)

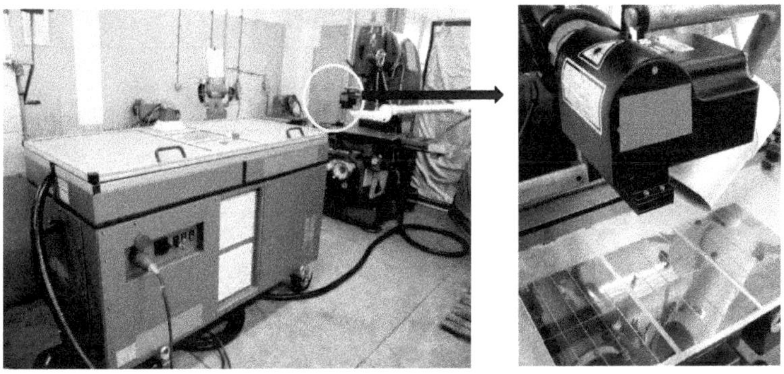

As described earlier, the laser action consists in excitation of the active medium with radiation from another source, and then releasing this energy in the form of a coherent radiation quantum (Zawieska, 2004).

A device with a laser, regardless of its type, contains basic elements: light source - laser, optical system for shaping the laser beam, optical system for guiding and observing the beam, device for mounting and moving the object, system for controlling the work of the laser and displacement device (Łunarski, et. al.,, 1989).

CHARACTERISTICS AND TYPES OF LASERS

The name LASER comes from the first letters of the words: **L**ight **A**mplification by **S**timulated **E**mission of **R**adiation.

Nowadays, a laser is a coherent electromagnetic radiation generator working in the spectrum from ultraviolet to far infrared.

Lasers can provide a beam with energy that exceeds the capabilities of many sources. In the Table 1 a comparison of the beam energy of different sources with the laser is presented [W•cm^{-2}] (Geleta, 2013).

Table 1. Beam energy of different sources (Geleta, 2013)

Energy source	Beam energy [W•cm^{-2}]
The sun	$5 \cdot 10^3$
Acetylene oxy-flame	10^4
Electric arc	10^5
Plasma	10^5
Electron arc	$4 \cdot 10^8$
Nuclear explosion	10^{10}
Laser CO_2 (continuous wave)	**10^9**
Nd:YAG laser (pulse wave)	**10^{14}**

Laser radiation is characterized by (Kusiński, 2000) characteristics:

- the beam is emitted in one direction defined by the axis of the optical resonator,
- the angle of divergence of the beam is small and amounts to several mrd,
- the distribution of radiation intensity in the cross-section, for the basic type of vibrations - that is the basic mode TEM_{00} (Transverse Electro-Magnetic) - is the distribution described by the Gauss function, for which the highest radiation intensity occurs on the axis (in the case of multimode operation this distribution is subject to modification),
- the beam can be focused with a simple optical system and produces a small diameter spot (10-300 μm) with increased power density,
- there is a possibility of continuous and impulse operation,
- high electric field strength of the electromagnetic wave (which is associated with high power density) may be accompanied by electrical discharge and ionization of its optical properties,

- lasers, depending on the type, can produce wavelengths of varying lengths (from 10 nm to over 1 mm, using the 200-1100 nm range for material processing),
- laser radiation is highly monochromatic,
- the radiation is coherent in time and space.

A typical laser device consists of four basic elements:

1. Active medium - laser action takes place there
2. Optical resonator - a system of mirrors that return the beam back to the ccntcr of thc tubc and amplify it (thcy also control thc wavclcngth of the emitted radiation)
3. A power source (pumping) that supplies electrons or photons (excites the laser)
4. Cooling system - collects heat from the laser medium and the resonator pipe

In addition, the laser systems used in the industry are equipped with:

5. Electrical power supply system - for laser and other actuators and control devices
6. Load-bearing structure and housing

Laser Division

There are several criteria for the division of lasers used in science. One of them is the division according to the type of active center.

Division of lasers by type of active medium (Elamed, 2019; Ziętek, 2009):

- **Gas:**
    - CO_2 laser (10.4 μm) - near infrared,
    - CO laser (5-6.5 μm) - near infrared,
    - N_2O laser (10.6 μm), molecular gas lasers CH_3OH, $C_2H_2F_2$, CH_3F (40 μm - 1 mm) - far infrared,
    - chemical lasers (in which one of the reaction products is: I, HF, HCl, HBr, CO, CO_2) emitting in near infrared (1.3-11 μm),
    - argon ion laser (476.5-514.5 nm) - visible range,
    - krypton laser (503.9-752.5 nm) - visible range,
    - helium-neon laser He-Ne (632,8 nm) - visible range,

  - excimer lasers such as ArF, XeCl, KrF, emitting 193, 248, 308 nm (UV) wavelengths,
  - nitrogen laser N_2 (337 nm) (UV).
- **Liquid**: used liquid lasers include dye lasers,
- **On a Solid Body:**
  - ruby laser (694.3 nm) - visible range,
  - Nd:YAG laser, in which the matrix is a yttrium-aluminum grenade, and the admixture is neodymium,
  - titanium sapphire laser (690-1080 nm) and other tunable solid-state lasers,
  - lasers on rare earth elements (holmium, erbium, thulium) emitting radiation of the wavelength of about 2 μm depending on the type of matrix and the type of admixture,
  - semiconductor (diode) lasers emitting in a wide spectrum range from visible to infrared.

Table 2 shows the use of different types of lasers and the differences between wavelength, pulse energy and operating mode.

Another classification method distinguishes lasers according to their operating mode (pulsed and continuous), power, pumping mode, emitted radiation wavelength, pulse duration and even harmfulness (Geleta, 2013; Major, 2002).

An illustration of the range of laser capabilities is shown in Figure 6 according to wavelength (Oczoś, 1988).

Characteristics of Laser Radiation

Laser radiation is characterized by monochromaticity, high temporal and spatial cohesion, directivity of beam propagation and the ability to obtain very high power densities. For these reasons, laser radiation is not easily dispersed and can travel long distances. The laser beam can be focused by optical systems up to several micrometers in diameter. This allows for a high power density of up to 10^{10} W/cm^2 (Oczoś, 1988) (see Table 2 above).

The effect of the laser beam on opaque surfaces is very characteristic and depends on (Łunarski et al., 1989):

- absorption of light flux and conversion of its energy into thermal vibrations of solid state networks,
- heating the material without destroying it,

Table 2. Application of lasers by type and wavelength (Geleta, 2013; Major, 2002)

Laser type	Wavelength	Energy	Mod	Application
On a solid body				
Nd:YAG	1.06 μm	1 W - 8 kW	Continuous or pulsed	Aluminum-itre garnet ($Y_3Al_5O_{12}$) doped with neodymium ions. Metalworking, medicine, measurements
Ruby	Red light 694 nm	MW	Impulsive	Pulsed holography, measurements
Semiconductor				
Semiconductor	Infrared to visible	1 mW - 100 mW	Continuous or pulsed	Optoelectronics
Liquid				
Dye	Continuous from infrared to visible	1 mW - 1 W	Continuous or pulsed	Medicine, measurements, optical spectroscopy
Gas				
CO_2	10.6 μm	1 W - 40 kW (100 MW in pulse mode)	Continuous or pulsed	Materials processing, medicine, isotope separation
He-Ne	632.8 nm or 1150 nm	1 mW - 1 W	Seamless	Measurements, holography
Argon	458 nm, 477 nm, 488 nm, 497 nm, 501 nm, 515 nm, and others	1 nW - 150 W	Continuous or pulsed	Printing houses, medicine, excitation of atoms in dye laser (liquid)
Excimer	198 nm, 248 nm, 308 nm, and others	1 kW - 100 MW	Continuous or pulsed	Micro engineering, medicine

Figure 6. The spectrum of electromagnetic radiation

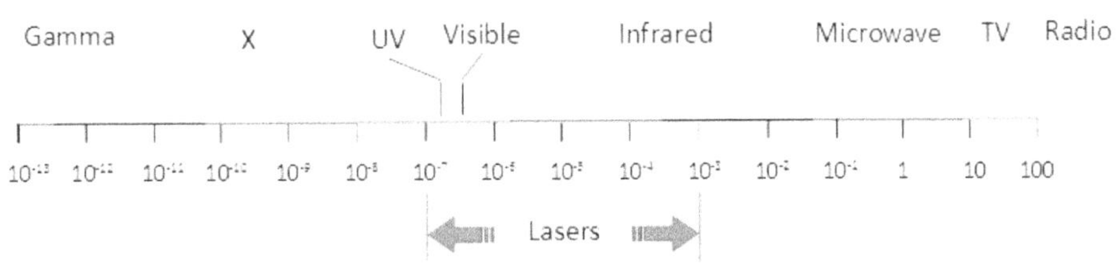

- development of evaporation of material in the area of the stream and dispersion of evaporation products,
- cooling of the material after the laser beam has stopped working.

With the pulse laser, individual craters are visible on the surface after the beam operation, while with the fixed laser, the trace is even.

LASER APPLICATIONS IN PRODUCTION: SELECTED APPLICATIONS

Laser Techniques Associated with Material Melting

In order to modify the properties of the surface layer with a laser beam, various methods are used, such as **strengthening by transformation, melting the surface layer** - during which the laser heats the thin layer of the surface being processed (e.g., surface hardening). Although the roughness of the surface deteriorates, the operational properties - tribological, fatigue and anticorrosive - improve. In this way, components made of many different materials can be modified: grey cast iron, tool steels, including high-speed cutting steels, stainless steels, bearing steels, carbon steels and low alloy steels, as well as aluminum alloys or pure titanium (Burakowski & Wierzchoń, 1995).

Another type of processing is **glazing**. Its aim is to obtain an amorphous structure in the top layer. It is quite a difficult process, because not all alloys are prone to amorphization. At high cooling rates, the viscosity of the liquid metal in some alloys increases so that the alloy does not crystallize. It solidifies in a disorderly manner, creating an amorphous alloy with the characteristics of glass mass. Glazing is successful in non-eutectic alloys with a tellurium matrix of Ag, Ga, Cu, In; with a lead and tin matrix (e.g., Pb-Sn, Pb-Au, Sn-Cu and others) and many others. Amorphous alloys are characterized by considerable strength and hardness, while maintaining ductility, although their strength at low temperatures decreases. They are characterized by high corrosion resistance, and some of them have special magnetic properties (Burakowski & Wierzchoń, 1995).

The melting of the surface layer also involves **laser compaction**, which involves melting the material to a certain depth in order to obtain a higher density of material. Most often, as a result of compaction, porosity, surface defects such as scratches, delamination, etc., are removed. Compacted surfaces of products are characterized by greater hardness, smoothness, higher resistance to tribological and fatigue wear, are more resistant to corrosion. Compaction is used to seal surfaces of, e.g., steel with carbides, heat sprayed coatings or galvanic coatings (Burakowski & Wierzchoń, 1995).

Similarly, when comparing machining parameters, **laser smoothing** is performed. The same phase and structural changes occur in the structure of the material, but the main objective is to reduce surface roughness and change the shape of the roughness profile. Sometimes the improvement in smoothness is achieved by bringing to the material melt zone (the so-called "pool") additives influencing the behavior of the material (the magnitude of turbulence, speed and direction of circulation), but then it is already an alloy smoothing process (Burakowski & Wierzchoń, 1995).

Laser Alloying

Laser alloying is the process of changing the chemical composition of the outer layer of the material surface through **the introduction of alloying elements**. The process consists in applying one or more alloying elements to specific zones of the workpiece surface, expected to modify the chemical composition of the surface layer. Then, by melting the input material and the material of the object (substrate), both of the materials are fused. The transition of the laser beam also triggers the self-hardening effect of the substrate material in the vicinity of the newly formed alloy (Łunarski et al., 1989). The products manufactured in this way are characterized by a surface layer of increased fatigue strength, better tribological and anticorrosive properties and less roughness. Non-metals (C, N, Si, B), metals (e.g., Co, Cr, Sn, Mn, Nb, W, Ta) and carbides of difficult-to-melt metals can be used as an alloying component or alloying components, while objects made of steel and cast iron are used for laser alloying. Alloying elements may take the form of pastes, powder suspensions in water or water glass, or coatings produced by other processes (film rolled, galvanically deposited). They are alloyed with single elements in order to improve heat resistance and high temperature resistance, anticorrosiveness and abrasion resistance (Dobrzański & Dobrzańska-Danikiewicz, 2019). The depth of melt can vary from 0.05-5 mm and the degree of saturation with the alloying element is obtained at the level from 0.1:1 to 0.81 to 1. The more complex the chemical composition of the substrate is, the smaller the effects of alloying are. Moreover, if easily fusible alloying elements are used, they can evaporate to a large extent, instead of moving into the alloy composition. The formed alloy layer is well bound to the substrate, does not delaminate, does not peel off and does not crack (Łunarski et al., 1989).

Continuous or pulsed lasers can be used for remelting, and the advantages of the laser melting process are (Łunarski et al., 1989):

- formation of solutions with higher solubility of alloying components than in traditional melting,
- economical use of alloying elements,
- small range of further heat treatment,
- high speed of the process, ease of its regulation, repeatability of results, high quality of the surface of the workpiece.

Laser Heat Treatment

Laser annealing, tempering and quenching are so-called heat treatment techniques not involving materials melting, used to induce changes in the structure of materials.

The main goal of **annealing** is to homogenize the structure. Laser beam splitting is reduced by structural defects, in some cases density may increase. Laser beam splitting is reduced by structural defects, in some cases density may increase. Since a laser dot is not large, laser annealing is used for local heating and thus for local reduction of hardness, plasticization, and subsequent deformation of these areas or increasing fatigue strength (Burakowski & Wierzchoń, 1995).

Laser tempering enables heating up to a certain temperature and cooling down, and is used to improve ductility and reduce brittleness. The local interaction of the beam offers a number of advantages: phase and structural changes take place in a similar way as in the furnace, but the carbides are more fine-grained, the material has higher hardness, strength, impact resistance. These properties can be adjusted by changing the operating parameters of the laser. In addition, stresses are locally removed, e.g,. hardening stresses (Burakowski & Wierzchoń, 1995).

The laser beam also enables **hardening,** which is used for machining tools, dies, parts of transport and agricultural machines. The laser beam also enables **hardening** which is used for machining tools, dies, parts of transport and agricultural machines (Łunarski et al., 1989). As a result of much higher heating and cooling speeds, it runs at a higher temperature than classical hardening. Laser-hardened materials increase fatigue strength, impact strength, toughness, ductility, corrosion resistance and abrasion resistance, obtain uniform hardness distribution and better surface smoothness (Dobrzański & Dobrzańska-Danikiewicz, 2019). In order to obtain these results, the treatment is carried out in such a way that the zones fortified by a single pulse overlap due to the specific shape of the energy distribution curve (reversed Gauss curve). The nature of changes in the properties of the processed material depends on the material itself, its thermal and physical properties, treatments prior to laser processing, as well as the size of the laser dot and the duration of the flux per unit of material volume (Łunarski et al., 1989; Radziejewska, 2011).

The use of lasers in heat treatment can bring measurable benefits. Laser can be used in industry not only as a monochrome light source, but also as a heat source with high power density. Laser heat treatment technology is increasingly being used in the production of parts worldwide. Laser is used to heat cast iron parts, nickel alloys, titanium alloys, cobalt alloys, copper alloys and aluminum alloys. Laser processing makes it possible to obtain the quality characteristics of the product surface layer that are comparable to, or even better quality than conventional furnace heating. It also has some additional advantages: the capability to process workpieces of any shape, weight, size, surface area or surface that is difficult to access. Laser hardening technology is much less energy-intensive than furnace hardening or induction hardening (Kusiński, 2000; Klimpel, 2012). The laser beam energy is only absorbed by the workpiece's surface layer and not by the workpiece as a whole up to the specified hardening temperature. In the case of laser hardening, the core remains raw. For this reason, the processing time is also much shorter than in traditional methods. During laser hardening process, the internal stresses are much smaller as a result of a considerably lower heat input. The object is therefore much less deformed and, as a result, finishing operations such as straightening or annealing can be dispensed with. This also saves material, as the machining allowance can be reduced. In addition, no hardening liquid (oil or water) is required for machining. You can harden the whole surface or its selected fragments with a laser.

The characteristics of (Klimpel, 2012; Colombo, 2003) are important for laser processing in connection with material heating:

- operating mode - continuous or pulsed,
- radiant power,
- wavelength,
- distribution of energy density in the cross-section of the emitted beam,
- shape and dimensions of the beam focus,
- beam speed and trajectory of movement (to ensure a homogeneous structure, pendulum movement, overlapping or pointless overlapping),

- type and flow rate of protective gas (argon).

In heat treatment processes, multi-mode, TEM_{mn}, continuous or pulsed lasers are used.

Laser Welding

Laser welding is a processing operation that uses consistent and monochromatic radiation emitted by laser to permanently bond parts together. Laser welding is a treatment process that uses consistent and monochromatic radiation emitted by a laser to permanently bond parts together. Laser radiation provides the most concentrated heat flux compared to other welding methods. This allows the elements to be connected with high accuracy and quality of the weld and results in less distortion. Laser welding is used for joining steel, aluminum and titanium alloys. Welding of aluminum alloys is difficult because the material has a low absorption capacity and a large part of the radiation is reflected from the surface of the object and dispersed in the environment. For this reason, high power lasers are used to weld low-absorbent materials. In addition, welding stations, generally specially designated, should have screens to protect the health of personnel.

Welding of titanium alloys, in turn, requires the use of gas shielding, and laser welding can be used to weld components up to a thickness of about 15 mm (Oczoś, 1988).

The power of a welding laser beam in gas lasers is adjusted by changing the power of the electric discharge system HF, in Nd:YAG lasers, fiber and disc lasers by changing the power of the pumping system of flash lamps, or diode pumping laser, in diode lasers of high power - by changing the intensity of direct current. With increasing laser power density during welding, the melting depth increases until the liquid metal area is penetrated and the metal leaks out. Sometimes exceeding a certain value of beam power density for a given material and its thickness can result in gas bubbles, cavities, uneven weld faces and metal leakage. In the process of welding with various techniques, the following parameters are also important: the focal dimensions of the laser beam, the depth of the focus and its position in relation to the upper surface of the welded joint (Klimpel, 2012).

Laser welding with a pulse or continuous beam and using a low or medium power laser requires, in addition to the gas blowing which removes the sheath of plasma from the welding area - also the protective gas blowing along the axis of the beam. For lasers up to 5 kW, argon, helium or a mixture of the two is used as a protective gas. In high-power CO_2 gas laser welding processes, it is recommended to blow a protective gas (helium) along the beam axis and, in addition, across the beam axis. Sometimes a double blow is recommended: at the top of the connector with inert gas, and above with an air stream that protects the laser optics. The type of protective gas depends on the type of material to be welded:

- for alloyed and high-alloyed steels - a mixture of argon+20-30% CO_2 or CO_2 only, with a flow rate of 10-12 l/min,
- low carbon steel - protective gas may be dispensed with if the sheet thickness <3mm, high welding speed is used 2-3 m/min. and the time during which the weld metal stays in oxidizing conditions is short,
- reactive metals (titanium and titanium alloys) - inert gas shielding is absolutely necessary.

In the laser welding process it is necessary to choose the right process parameters owing to specific phenomena. For example, it is advantageous to weld aluminum alloys using low power, which makes it possible to obtain a narrow weld with favorable mechanical properties (better than conventional TIG arc welding). Aluminum alloys are difficult to weld because their absorption coefficient of laser radiation is low. A laser beam reflects intensely on the material and welding is only possible after the formation of a weld pool. Sometimes objects are covered with an absorbent substance, they are anodized or their surface roughness is increased by mechanical or chemical treatment.

Copper and its alloys have a similarly low absorption coefficient. In addition, a high beam power density is required due to high thermal conductivity.

Moreover, a double-beam technique is used for welding aluminum, copper and magnesium. The material is welded with two beams, one melts the surface of the low-absorbent material and the other forms a keyhole or pool of liquid metal weld. During this process, the power distribution of both beams is controlled to ensure proper cooling conditions.

Modern laser welding also makes it possible to combine plastics. This is an advantagcous solution bccausc of thc widcsprcad usc of plastics in structures. Plastic welding, contactless and direct, melts the material, and after it has solidified, a permanent connection is achieved. The welding of transparent plastics, for which the beam absorption coefficient = 0, can be problematic. In this case, substances absorbing laser energy are introduced into the melting zone. The welding speed (linear energy) should also be adjusted to avoid pyrolysis, carbonization or decomposition of the polymer.

In the process of laser welding the parts to be joined should match, the edges should be clean. Too large a gap may cause a defective connection, with defects such as: concave face, lack of melting, sticking of the material. Items are generally cleaned at the point of welding (there is no need to clean the entire item) mechanically or chemically. Moisture, which can cause gas bubbles or crystalline and hydrogen cracks, is also avoided. Mutual movement of parts is less important.

Comparing laser welding with other traditional methods, it is:

- highly efficient, stable,
- easy to robotize and automate,
- enables considerable depths of melting into the material to be achieved,
- the resulting connector is characterized by high mechanical properties,
- minimum susceptibility to defects,
- narrow heat affected zone.

Laser Cutting

Cutting and cutting out elements are performed by local melting and evaporation of the material by means of a laser beam of a specific energy. Laser cutting technology can be classified into the following categories:

- cutting thin or thick layers of material without damaging the substrate on which they are applied - used in electronics,
- shallow cutting of materials, e.g., ceramics or semiconductors (application as above),
- cutting of metals and other materials of greater thickness, used in the production of various products.

CO_2 lasers with continuous and high power work are used to cut materials. They can be used to cut steel, titanium and its alloys, tantalum, zirconium and its alloys, nickel and aluminum alloys, as well as non-metals - such as wood, plastics, textiles, leather, etc. (Powel J. 1998).

Laser cutting is characterized by high cutting speed, easy cutting of complicated shapes, relatively good quality of cutting edge, small size of thermal interaction zone in the material being cut and contactless operation (Nowicki M. 1978).

Cutting operations, depending on the material being processed, can vary in speed, while laser beam parameters are important with respect to the thickness of the material being cut, the width of the cut and the size of the heat affected zone. In many cases cutting should be performed in an atmosphere of active gas such as oxygen or inert gas such as argon or nitrogen. The active atmosphere is most often used for metalworking, gas significantly increases the absorption of laser radiation due to oxidation of the cut metal surface, increases the surface temperature at the processing site by releasing heat in the oxidation reaction, removes fumes and fragments of molten metal from the processing site and facilitates access of the laser beam to the cutting slot. The use of oxygen increases cutting speed by approx. 40%. The use of oxygen increases cutting speed by approximately 40%. However, it may degrade the quality of the cut edge and increase the size of the gap. In each case, the laser operating parameters and oxygen flow should be selected according to the type of material in order to avoid the above mentioned difficulties, but also not to excessively cool the material by needlessly excessive gas flow. Some materials, such as gold, have to be prepared for cutting and are only treated as thin layers on a dielectric substrate (to avoid burning of the material). and undergo treatment only as thin layers on a dielectric substrate (to avoid burning of the material). Cutting efficiency is also increased by an inert gas. This is due to the fact that molten or evaporating material is removed from the processing area, the beam access is facilitated and shifted into the kerf, and the upper edges of the kerf are prevented from burning as a result of gas cooling. In an inert gas atmosphere, metal, wood, plywood, plywood, cardboard, paper, asbestos, textiles, rubber, plastics, ceramics and many others can be cut. All cutting operations must be controlled in terms of appropriate parameters selection, i.e., cutting speed, laser beam power, beam focus, gas stream magnitude, as they determine the depth, width and shape of the cut (Nowicki, 1978).

Sometimes gas is not needed, e.g., in modern production solutions using low power lasers (20-35W), the possibility of cutting materials (an example is cutting aluminum tiles up to 1.5 mm thick). Another way is to use fiber and diode lasers (especially high power - the so-called multi kW) to cut elements of considerable thickness - over 10 mm - then cutting can be done in a nitrogen atmosphere (Laser, 2012) (Figure 7 and 8).

In the context of plastics cutting, the use of a laser beam can be beneficial for many reasons. Plastics are not only subject to plastic processing but also to machining. In this case, laser cutting makes it possible to avoid difficulties, especially with poor surface quality. The product quality is particularly

Figure 7. Thin-walled decorative element made of aluminum alloy (a), composite (b)

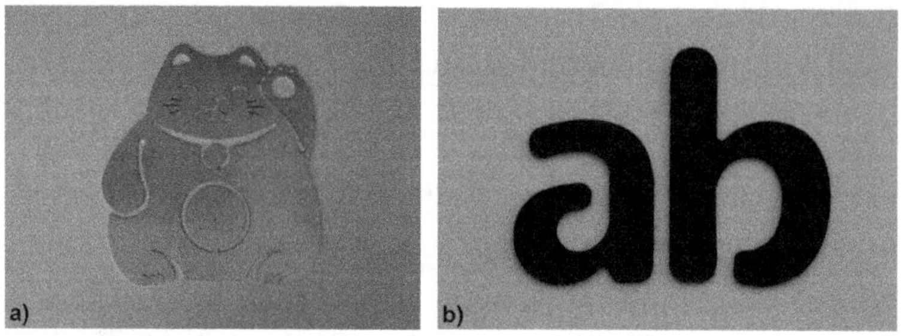

Figure 8. Products of considerable thickness cut by laser

dependent on the chip formation process. The machining of polymers can produce various kinds of chips: continuous, discontinuous, plastic deformed, etc. Continuous shavings can make machining difficult because of their tendency to swell or to stick to the workpiece or tool, impairing not only the visibility of the machined area but also the work safety. The build-up of chips can increase the cutting forces and raise the temperature, making it easier for the polymer to chip. Similarly, discontinuous shavings - torn off (chipped off) significantly worsen the condition of the machined surface. A special case of discontinuous chips is fine dust, which also causes faster tool wear (Darlewski, 1978).

Plastics are a group of modern materials used in various fields of production. The characteristic properties of plastics mean that they are not only substitutes for metal materials, but also make it possible to design completely new products often with surprising shapes and functional properties. In comparison with metals, they have a number of advantages, such as low density, vibration damping, electrical insulation or controlled electrical conductivity, chemical resistance, and corrosion resistance and, due to the various additives used, the ability to modify properties (Darlewski, 1978).

An alternative way of plastic cutting, with regard to the earlier presented mechanical cutting (machining), can be laser beam cutting which has become increasingly popular in recent years. Laser cutting is a direct action of the laser beam on a selected area of material. In order to heat the material in the cutting slot area to a melting or evaporation temperature, the laser beam must provide the required energy (Klimpel, 2012; Šuba et al., 2018). In most cases, the material is immediately evaporated and the heat affected zone is very small.

Laser cutting is used to cut most engineering materials: metals, ceramics, plastics, wood, and paper. The cutting process uses single mode, $TEM_{00,}$ continuous or pulsed lasers. They make it possible to focus the beam to obtain the smallest diameter and the greatest depth of the beam. This is essential to ensure high quality cutting (Powel, 1998).

To determine the optimum cutting parameters the following aspects are considered (Powel, 1998):

- laser beam power,
- operating mode (continuous or pulsed) and wavelength,
- speed of beam movement on the cutting track,
- diameter and depth of focus of the laser beam,
- focal length of the beam,
- position of the focus of the beam in relation to the upper surface of the cut object,
- type and pressure of the associated gas,
- the diameter of the associated gas nozzle and its distance from the surface of the workpiece,
- thickness of the material to be cut.

In the case of laser cutting, the focus of the beam is important. It should be kept at a constant distance along the entire length of the cut. The stability of the focus position is ensured by control systems and executive mechanisms of laser devices or robots. Defective location of the focus, its changes during the processing of one product may result in excessive rounding of edges, or in extreme cases - lack of cut.

One of the main advantages of laser cutting is the high speed of the process. It is only limited by the power of the beam and the parameters of the system guiding the laser head along the cutting line. The cutting speed depends on the power density of the beam and the thickness of the material being cut, as well as the type and pressure of the accompanying gas. When cutting iron-based materials with oxygen, a small change in the chemical composition affects the optimum cutting speed. Material cut at too low a speed can have damaged edges. When cut at too high a speed, it can have an edge with metal and slag infiltrations in the lower part. The accumulation of metal and slag can in extreme cases lead to a lack of cut across the entire thickness.

The quality of cutting plays a special role in, e.g., profiled cutting of metal sheets, when problems may appear in corners, bends, then computer control and modification of the motion path of the head is applied so that the beam does not slow down or stop at a point (it causes burning, excessive melting of the material and, as a result, the appearance of defects) (Klimpel, 2012).

Examples of cutting parameters singled out for selected materials are given in the Table 3.

Table 3. Examples of cutting parameters of steel sheets (Klimpel, 2012)

Feature	Carbon steel sheet, thickness 40 mm	Carbon steel sheet, thickness 15 mm	Sheet of austenitic steel, thickness 10 mm
Laser	CO2, beam power 2, kW	Fiber, beam power 1.5, kW	Fiber, beam power 4, kW
Focal length diameter, mm	4	0.3	0.3
Cutting speed, mm/min.	0.4	1	1
Location of the beam focus	On the upper surface of the sheet metal	7 mm below the upper surface of the sheet metal	10 mm below the upper surface of the sheet metal
Accompanying gas, pressure, bar	Oxygen, pressure 6-8	Oxygen, pressure 1	Nitrogen, pressure 20
Nozzle diameter, mm	3	1.5	2.5
Distance of the nozzle from the surface, mm	4-5	1	0.7

The laser cutting process requires a simultaneous, coaxial flow of gas, called the associated gas, with the laser beam. It blows liquid and evaporated or molten material out of the cutting slot. The accompanying gas may be chemically inert - argon or nitrogen, or reactive - oxygen, air or other oxygen mixtures. If there is a reactive gas in the iron matrix, liquid metal and combustion products are thrown out of the cutting slot. During cutting plastics and wood, these are only combustion products (Klimpel, 2012).

Oxygen is recommended for cutting non-alloyed steels and alloy steels, while nitrogen or argon is recommended for high-alloyed steels. The quality of the cut gap, especially the smoothness and removal of the overhang from the material, also depends on the pressure of the accompanying gas and the stability of its flow (without turbulence or turbulence).

As the laser beam moves along the cutting line, liquid metal, polymer and ceramics flow in the opposite direction, while the accompanying gas removes the liquid material from the gap before it solidifies. A very thin layer of solidified material remains on the cut edges. In addition, the gas protects the laser optics against contamination by splashing material or vapors.

The gases used in laser cutting processing and the materials to which they are dedicated are listed in the Table 4.

Table 4. Accompanying materials and gases (Klimpel, 2012)

Gas	Materials
Air	Aluminum, plastics, wood, composite materials, aluminum oxide, glass, quartz, rubber
Oxygen	Carbon steels, low alloy steels, corrosion-resistant steels, copper
Nitrogen	Corrosion-resistant steels, aluminum, nickel alloys
Argon	Special steels, high-alloy steels, nickel alloys, aluminum and aluminum alloys, titanium, tantalum, zirconium

Depending on the material to be cut, the pressure of the accompanying gas is also selected. Usually the pressure is from 1 to over 25 bars. Pressure has a key influence on the formation of material overhangs at the lower edge of the cut.

In addition, when cutting, the distance between the nozzle and the surface of the material is chosen. This parameter affects the width of the intersection slot. One of the advantages of laser cutting is the narrow gap. Too large a distance between the nozzle and the nozzle results in an extended gas stream. It is no longer as effective as the narrow stream, especially in the lower areas of the gap, and the quality of the cut surface also decreases. Incorrectly placed nozzles - i.e., too far or at the wrong angle (they should be coaxial with the beam) are the cause of disturbances in the flow of liquid material, turbulence. The diameter of the gas nozzle is also selected in order to ensure proper cutting quality and economical gas consumption (Klimpel, 2012).

Laser cutting process is nowadays one of the basic technologies, providing high accuracy and quality of cut edges. Laser cutting has many advantages:

In Technical Terms

- less material deformation and lower internal stresses,
- narrow heat affected zone (minimum heat impact on the material),
- no machining (which is necessary after plasma or oxygen cutting),
- non-contact machining and no mechanical damage.

In Terms Of Processing Result

- high dimensional accuracy and edge quality, sometimes exceeding the diamond cutting quality,
- favorable condition of the structure and the edge of the intersection.

In Economic Terms

- high cutting speeds can be used,
- ease of computerized process control, automation and robotization.

In Environmental Terms

- less emission of harmful fumes and dusts than when cutting with oxygen or plasma,
- no waste typical of conventional cutting.

Laser cutting, however, has certain limitations:

- thickness of the elements to be cut (usually up to 25 mm),
- sometimes curvature, distortion of the product,
- emission of invisible electromagnetic radiation for the operator,
- high cost of high-power equipment.

In the laser cutting process, polarization of the beam is important, on which the cutting efficiency depends. The highest efficiency is achieved with linear and elliptical polarization and the best cutting quality is achieved when the polarization direction is in line with the cutting direction.

Slightly worse process parameters are obtained with circular and random polarization.

Circular polarization ensures uniform quality in all directions, but the cutting efficiency can be reduced by about 40%. This treatment is recommended for cutting metals with oxygen as a companion gas. The use of oxygen (or air or other mixture containing oxygen) initiates an additional exothermic reaction of burning liquid metal in the cutting slot. This increases the heat energy of the process. Less efficiency is then an advantageous solution.

The laser cutting process is influenced by a number of factors:

- **Laser Device**: Power, laser beam quality, guiding and shape of the beam, cross-section, polarization,

- **Adjustable Parameters**: Beam power, mode, beam focus and focus, beam diameter, cutting speed, working or protective gas, its composition and pressure,
- **Object**: Material (thermal conductivity, structure, mechanical properties) and geometry (dimensions, shape, geometric structure of the surface).

Laser Hole-Drilling

Laser hole drilling is a commonly used variant of laser cavity processing. Treatment consists in removing locally melted and evaporated material under the influence of a focused laser beam with high power density (Oczoś, 1988).

Laser drilling has many advantages:

- the machining is non-contact, therefore the tool is not damaged, worn or heated,
- because the beam is optically oriented, the hollow accuracy is very high,
- there are no chips typical of drilling and milling, the material is removed by ablation (melted and evaporated),
- no burrs, holes are highly accurate and the edge is of high quality, no finishing is required,
- laser allows to make long holes with small diameters,
- it is possible to make holes in difficult to machine materials (diamond, aluminum oxide, tungsten, titanium),
- holes may be drilled in hard-to-reach places and at a slight angle (10°),
- the treatment is safe in the environmental context - no substances characteristic of electrochemical drilling are used,
- in a single setting you can not only drill holes, but also cut and weld, and the process can be fully computerized, automated and robotized.

The reproducibility of the hollowing process is dependent on variations in laser power and radiation intensity (beam quality). The problem is that the removed material is deposited on the optical system used to focus the beam. A stream of compressed air, directed perpendicularly to the beam, is also used for hollowing. The protective plate of the machining optics system is then protected against melted material. In this way, the diameter of the beam must not be dispersed or enlarged.

Laser drilling can be carried out with different types of lasers:

- **on the solid body**: **neodymium** Ng:YAG (for drilling holes in metals with a diameter of more than 5 μm, with a frequency of pulse repetition up to 300 Hz, and power up to 400W) and **ruby** (it is especially suitable for drilling thin films, diameters even from 0.3 μm),
- **gas**: **molecular CO_2** - used for processing dielectric materials (glass, ceramics, plastics) that almost completely absorb laser radiation, used in continuous operation with a power of 100 to 200 W; and **excimer lasers** - used for micro-drilling holes with diameters <500 μm, in hard and brittle materials (ceramics, glass) with a thickness of up to 1 mm (due to extended process time).

In the case of pulse laser processing, the drilling process and its result are crucial for the course and outcome of the hollow process:

- power,
- pulse duration,
- mode type and divergence,
- pulse frequency.

A hole is formed by a beam that is focused on the surface of the object, with a specified power density. In addition, the mechanical and thermal properties of the material, the pressure and the type of protective gas influence the result of the excavation.

Variable radiations intensities are used to produce holes that are free from burrs and burrs and do not require finishing. As the depth increases, the power density increases. This is related to the phenomena occurring during drilling. The constant value of the radiation intensity causes solidified material to settle on the wall of the hole as the process progresses. This can lead to the hole being closed again and opened again.

Marking

Laser marking is one of the varieties of cavity processing. Laser marking consists in making inscriptions or images on finished products or components. The signs are durable and do not require any additional processing. In many production applications, it is preferred for its advantages (Sobotova & Badida, 2017):

- flexibility - expressed in the ability to program and execute individual and differentiated graphic symbols in a single operation,
- ability to mark most materials,
- marking in hard-to-reach places,
- miniature marking,
- high repeatability of the position of the characters,
- significant time savings compared to mechanical engraving,
- high process speed,
- marking of any surface - flat, cylindrical, mirror inscriptions, negatives, arches, etc.

Laser marking is the impact of a beam on a material, its melting, evaporation or photochemical reaction. In this way, monochromatic or colored marking is possible.

Figure 9. Business cards monochromatic (left) and color (right) made with laser

Marking can be performed by a focused beam or by a mask.

Focused beam marking has two varieties: dot laser marking or laser engraving. Point marking is performed by a pulse laser whose control system allows it to work with a low pulse frequency (100 Hz or less). This results in gaps between the marks created by the evaporation of the material from the surface of the object. The marking stations are now computerized, allowing you to define the processing location, mark placement and character depth. Continuous marking is performed with a continuous laser with forced impulse work (10 Hz and more). The marks on the surface of the material overlap to form a continuous line. This method of marking is more flexible compared to point marking, it facilitates characters reading using electronic readers as well as makes it easier to create small characters (Sobotova & Badida, 2017).

Mask marking should be performed with a laser that produces a large diameter beam. This requirement results from the necessity to cover the template surface with a beam. One impulse is enough to make a graphic symbol. If the mark is too large, or there are difficulties in using a laser with a sufficiently large diameter, multiple pulses are applied, moving the workpiece in relation to the head in order to cover the entire mask step by step. Mask marking is often used to mark moving objects, e.g., on a production line, because the pulse is short and the travel speed low, which does not cause smudging, marking or symbol distortion. Many lasers are capable of producing 50 pulses per second.

Laser marking is a process that can be used on an extremely large number of materials. It marks products made of metals and their alloys, plastics, glass, ceramics, leather, rubber, wood, paper and even food (egg shells, apples, and others). Special additives are also used, especially for plastics that change their structure under the influence of laser radiation. This treatment produces marks of a different color or visible under the influence of UV radiation.

Laser Strengthening

In the technology of manufacturing components of high strength, especially fatigue strength of elements subjected to changing loads, there is a method known as crumpling. It has its own variations - static and dynamic. Dynamic methods include crushing with a stream of a specific medium, e.g., sand-blasting, used more often to remove unnecessary layers, e.g. varnish or corrosion, but also shot-peening, used as a strengthening treatment (Krasucki & Majewski, 2010).

The impact of the ball stream results in the creation of favorable compressive stresses in the surface layer of the product, which translates, among others, into an increase in hardness and fatigue strength. Such processing takes place in special chambers, to which compressed air is supplied. The chamber is equipped with a system of special nozzles which suck in the balls and throw them in the direction of the product by means of an air stream. Often the chambers are equipped with mechanisms to move parts in order to ensure that the surface is evenly covered with machined traces (in the case of metals of low hardness, they are visible in the form of dents created by balls), and thus achieve a proper state of stress. The balls used in the machining process, however, wear out over time: they break, deform, and then the entire batch should be replaced with a new portion (Hammersley et al., 2000).

The existing strengthening technology can be replaced by a more modern method - laser shot peening (LSP). The machining of parts is based on the impact of a short-lived, high-energy laser pulse. A shock wave with very high pressure amplitude is generated and, as a result, plastic deformations are generated on the metal surface that cause compression stresses. The resulting technological surface layer shows increased wear and corrosion resistance, resistance to cracking and has a higher fatigue strength than the native material - just as it does after traditional crushing (Krasucki & Majewski, 2010). Using new generation lasers with high energy density (from several dozen to 300 and more J/cm^2) and very short pulse duration (less than 50 ns), the height of radiation power density of 106-1011 W/cm^2 is obtained. Thanks to this, shock waves of very high pressure amplitude (up to 10 GPa) can be easily produced in solids (Hammersley et al., 2000).

Obtaining such high values of pressure is possible through the application of some additional technological solutions. For LSP, the following procedures are implemented: the treated surface is precoated with a thin absorption layer of thickness ranging from a few to nearly twenty µm and an additional inertia layer several mm thin is applied. The absorption layer can be a black paint coating. Its purpose is to prevent the penetration of heat into the metal. It also protects the metal from the destructive effects of the laser beam (ablation, i.e., evaporation). Then, the inertial layer can be a glass plate, a layer of water, or another transparent to radiation dielectric. Its purpose is to limit the rate of expansion of the plasma produced, so that its pressure (plasma) increases rapidly (Yongkang et al., 2013). The laser enhancement mechanism is as follows: the radiation absorbed by the absorption layer heats it up to approx. 10,000°C, ionizes it and produces plasma. During this process, a shock wave is generated from a very high pressure. The inertial layer limits the speed of the plasma movement, which increases the plasma pressure. The impact wave (high pressure) creates variable deformations in the material. The final effect is a plastic deformation, which causes residual compressive stress in the upper layer. This is a more energy-efficient and environmentally friendly machining method than traditional shot-peening with the use of balls (Peyre et al., 2000).

Laser Additive Techniques

Overlay Welding and Metallization

Overlay welding and metallization - these are the processes of applying various metallic layers to previously shaped surfaces. The purpose of the treatment may be to regenerate or repair a part by restoring the geometry of the product which has been lost in service (e.g., worn out as a result of friction); or to apply a coating having certain characteristics different from the original characteristics of the product (Oczoś, 1988; Oczoś, 1988; Chua et al., 2010).

Welding is also used to regenerate worn-out parts. Then, the melted metal is used to fill cracks, crushes and it can recreate worn out surfaces. In addition to the restoration of the shape and lost properties, it is often possible to improve the properties (surface refining). This is made possible by the application of a modified material composition. In this way, a surface layer is formed that is more resistant to corrosion, erosion, abrasive wear, cavitation, and with increased heat resistance and creep resistance. Overlay welding is performed on steel, cobalt alloys, titanium alloys, aluminum alloys (Blicharski, 2009).

3D Printing

The above mentioned ways of using the energy emitted by lasers do not exhaust the areas of their application. Other machining methods comprise the production of thin and hard coatings, including coatings with special properties, or intensive incremental shaping developed in recent years, including 3D printing, selective laser sintering (SLS) and others (Burakowski & Wierzchoń, 1995; Budzik, 2007).

Incremental shaping, also called additive fabrication. Freeform fabrication, solid freeform fabrication, digital fabrication, rapid prototyping), includes methods and technologies based on layered and fast manufacturing of products. Incremental shaping, also called additive fabrication (freeform fabrication, solid freeform fabrication, digital fabrication, rapid prototyping), includes methods and technologies based on layered and fast product manufacturing. The machining process uses a product geometry modeled as 3D-CAD (three dimensional structures in Computer Aided Design software), which makes it possible to manufacture products with almost unlimited geometric complexity and for a variety of purposes. The product is created very quickly (rapid) in comparison with conventional production. The creation of a product consists in combining loose or liquid material in a chosen way (incrementally) and is used in the production of not only models and prototypes (Rapid Prototyping), but also tools (Rapid Tooling) and finished products, both individually and in serial production (Rapid Manufacturing).

Figure 10. Laser additive methods

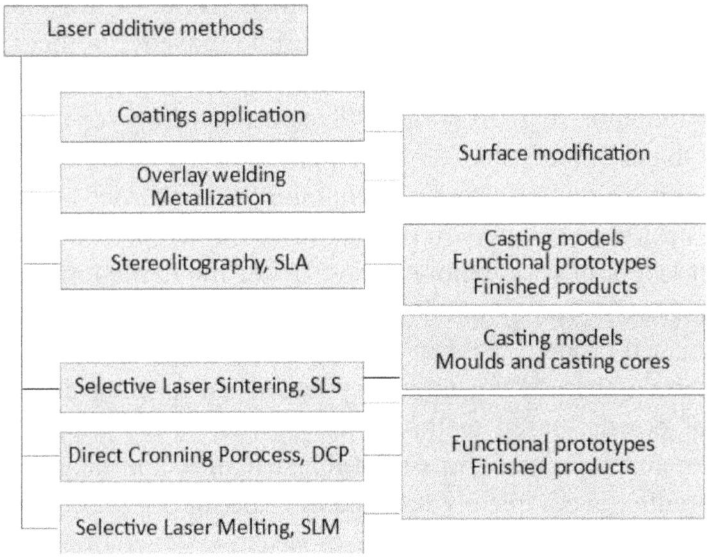

SLM - Selective Laser Melting: a method consisting in joining powder grains by melting them (e.g. SLM of MTT Technologies, LaserCusing of Concept Laser, or LENS of Optomec). Any material can be used in this method provided it is in powder form: titanium and its alloys, stainless steels, tool steels, zinc and its alloys, copper and its alloys, silicon carbide, aluminum oxides. As a result of powder grains melting there are no pores, gaps between grains. High-power lasers are used. The advantage of SLM products are their mechanical and operational properties similar to those of products made with the use of cavity techniques. Objects after being subjected to SLM need only finishing, e.g., sanding, in order to achieve a smooth surface. The SLM method is used to produce machine parts, molds, die-cutters, implants (automotive, electronics, aerospace and aerospace industry, medicine and dentistry).

SLS - Selective Laser Sintering: is a method of manufacturing products from metal powders or plastics (e.g. Direct Metal Laser Sintering of EOS - products, Direct Croning process of molding sand for production of foundry molds and cores). The powder grains are melted over the surface. Powders must be prepared before processing - they are covered with special substances that act as binders. This substance melts under the influence of a laser beam and thus merges the powder grains. In order to prevent the formation of gaps between grains, copper alloys are used as fillers that increase the density of the element. The source of radiation in this method is a CO_2 laser or Nd:YAG.

SLA – stereolithography: is a method based on local, layered photopolymerization of a liquid nanomer (nanomers - polyolefin resins) using laser radiation (e.g., SLA of 3D-Systems company) or UV light (e.g., Polyjet of Object Geometries Company). The light source - laser - is equipped with a beam positioning system; the container in the device contains liquid resin. The next layer is cured on the platform, the scraper smoothes out the subsequent layers of resin before curing. After the hardening of the layer, the platform is reduced by one layer. The product is formed layer by layer and the finished product is taken out of the liquid resin. As the SLA uses liquid materials, there is no support system when forming the product (in the case of powder-based methods, the powder is the support). Special support structures are made to stiffen the structure, and they are removed after the product is completely hardened. Sometimes a product made with the use of a laser beam is additionally exposed to UV radiation. This method requires a finishing treatment (removal of supports, improvement of surface smoothness). It is time-consuming and labor-intensive, limited only to photo-curable materials. It is also quite expensive, and the mechanical properties of products can deteriorate over time. It is used to manufacture small products, jewelry, implants, and molds for casting.

Additive methods make it possible to manufacture products with very complex shapes, sometimes impossible to obtain by machining in the form of a monolith (then a combination of semi-finished components was used).

Such processing is much faster than traditional methods, it is waste free and energy efficient. The spatial model of the product in the form of CAD files is loaded by the specialized computer software of the device. The laser control program is then generated. The laser beam interacts in a strictly defined place, reproducing a given geometry of the product in the material. Therefore, there are practically no restrictions as to the shape of the products.

However, it has its limitations: in the case of materials susceptible to oxygen, a protective gas is required that is individually matched to the type

Figure 11. Laser beam 3D printing: a) 3D printing process is controlled by computer, b) laser beam positioned above the platform draws a layer for the design, c) laser bonds the resin layer by layer, d) the printing is finished – the platform rises up, e) cleaning and removing resin, f) final product

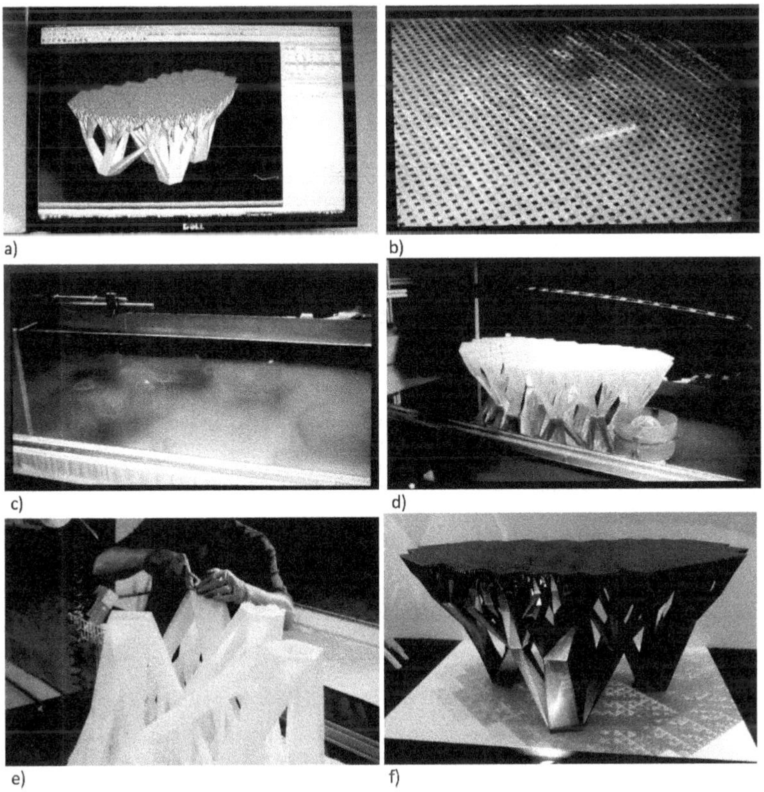

of material. The quality of the product - layer thickness, surface roughness - depends on the size of the laser dot (Siemiński & Budzik, 2015).

ADVANTAGES AND DISADVANTAGES OF LASER PROCESSING

Compared to other conventional processing methods, laser beam processing has the following advantages (Geleta, 2013):

- non-contact machining,

- no wear of the tool in the classic sense (e.g., dulling of the blade in machining),
- clean machining - no chips are formed, no machining liquid is needed,
- quiet machining,
- the laser beam can be separated and separately passed through a series of mirrors,
- the heat affected area is minimum,
- the technological process can be automated, robotized, the laser device can be integrated into a sequence of operations in a flexible manufacturing system,
- control over energy portions,
- toxic substances resulting from the treatment of certain materials can be separated (removed by industrial ventilation systems).

In addition, the use of lasers for surface treatment of machine parts makes it possible to improve the functional properties of the surface layer. These possibilities are beneficial because (Łunarski et al., 1989):

- only local changes in the condition of the surface layer are possible and the hardness is 15-20% higher than in the case of traditional methods of processing,
- processing in hard-to-reach areas is possible, the laser beam can be guided by an appropriately directed optical system,
- it is possible to change the chemical composition only in the near-surface area of the object,
- the beam impact zones are discontinuous, then the top layer is neither brittle nor prone to cracking.

Disadvantages (Geleta, 2013):

- so far quite a high price of good quality lasers,
- requirement for qualified personnel,
- necessity to meet safety requirements,
- occasionally low process efficiency, 0.2 - 15%.

CONCLUSION

The laser beam machining tool described in this chapter is not a new development, however, its constant modernization and popularization may become an interesting alternative to many different technological operations carried out by modern industry. Attention should be drawn to technological capabilities of lasers and their particular features, such as the possibility of correlation of radiation parameters with the purpose of processing, durability, repeatability, the possibility of local processing with a small heat-affected zone, and many others, which provide sufficient basis for more consideration of changing conventional machine tools and stands to modern laser devices. In addition, laser processing is generally waste-free (or low-waste, as in the case of, e.g. sheet metal cutting, where it is possible to optimize the position of the cutout elements on the sheets and, as a result, minimize waste). However, the problematic issues concerning lasers must be taken into account, e,g. noise - laser equipment is usually quiet, yet noise can be generated by using additional equipment necessary to remove treatment products or emissions of the toxic substances to the atmosphere as a result of interaction between a concentrated energy beam and the surface of the material.

REFERENCES

Blicharski, M. (2009). Inżynieria powierzchni [Surface engineering]. Warszawa: WNT.

Budzik, G. (2007). Synteza i analiza metod projektowania i wytwarzania prototypów elementów o skomplikowanych kształtach na przykładzie wirników turbosprężarek [Synthesis and analysis of design methods and prototype production of complex shapes on the example of turbocharger rotors]. Rzeszów: Oficyna Wydawnicza Politechniki Rzeszowskiej.

Burakowski, T., & Wierzchoń, T. (1995). Inżynieria powierzchni metali [Metal surface engineering]. Warszawa: WNT.

Chua, C. K., Lcong, K. F., & Lim, C. S. (2010). *Rapid Prototyping – principles and application* (3rd ed.). Word Scientific Publishing. doi:10.1142/6665

Colombo, V., Mentrelli, A., & Trombetti, T. (2003). Time-dependent 3-D modeling of laser Surface heating for the hardening of metallic materials. *The European Physical Journal D, 27*(3), 239–246. doi:10.1140/epjd/e2003-00281-7

Darlewski, J. (1978). *Obróbka skrawaniem tworzyw sztucznych warstwowych* [Machining of laminated plastics]. Warszawa, WNT.

Dobrzański, L. A., & Dobrzańska-Danikiewicz, A. D. (2019). *Obróbka powierzchni materiałów inżynierskich* [Surface treatment of engineering materials]. Retrieved from http://www.openaccesslibrary.com/vol05/5.pdf

Elamed. (2019) Retrieved from https://opm.elamed.pl/artykul/bezpieczne-zastosowanie-laserow/42545

Gebhardt, W. (2007). Rapid Prototyping. München-Vien: Carl Hanser Verlag.

Geleta, V. (2013). *Progresivne technologie obrabania* [Progressive fluid technologies]. Bratislava: Slovenska Technicka Univerzita v Bratislavie.

Hmmersley, G., Hackel, L. A., & Harris, F. (2000). Surface prestressing to improve fatigue strength of components by laser shot peening. *Optics and Lasers in Engineering, 34*(4-6), 327–337. doi:10.1016/S0143-8166(00)00083-X

Klimpel, A. (2012). Technologie laserowe [Laser technologies]. Gliwice: Wydawnictwo Politechniki Śląskiej.

Klimpel, A. (2012). Theoretical basis of laser cutting of metals. *Przegląd Spawalnictwa, 6*, 2–7.

Krasucki, J., & Majewski, M. (2010). Identyfikacja charakterystyk warstwy wierzchniej w procesie udarowego umacniania laserowego LSP [Identification of surface layer characteristics in the LSP laser impact strengthening process]. *Przegląd Mechaniczny, 7-8*, 45–51.

Kusiński, J. (2000). *Lasery i ich zastosowanie w inżynierii materiałowej* [Lasers and their application in material engineering]. Kraków: Wydawnictwo Naukowe AKAPIT.

Łukasik, Z., Kuśmińska-Fijałkowska, A., & Żurek, M. (2016). Bezpieczeństwo podczas pracy z laserami i urządzeniami laserowymi [Safety when working with lasers and laser devices]. *Autobusy, 10.*

Łunarski, J. (1989). *Obróbka powierzchniowa* [Surface treatment]. Rzeszów: Wydawnictwo Politechniki Rzeszowskiej.

Major, B. (2002). Ablacja i osadzanie laserem impulsowym [Pulse laser ablation and deposition]. Kraków: Wydawnictwo Naukowe AKAPIT.

Nowicki, M. (1978). Lasery w technologii elektronowej i obróbce materiałów [Lasers in electron technology and material processing]. Warszawa: WNT.

Oczoś, K. (1988). Kształtowanie materiałów skoncentrowanymi strumieniami energii [Shaping materials with concentrated energy streams]. Rzeszów: Wydawnictwo Politechniki Rzeszowskiej

Oczoś, K. (1988). Kształtowanie materiałów skoncentrowanymi strumieniami energii [Shaping materials with concentrated energy streams]. Rzeszów: Wydawnictwo Politechniki Rzeszowskiej.

Peyre, P., Scherpereel, X., Berthe, L., Carboni, C., Faabro, R., Beeranger, G., & Lemaitre, C. (2000). Surface modifications induced in 316L steel by laser peening and shot-peening. Influence on piting corrosion. *Materials Science and Engineering A*, *280*(2), 294–302. doi:10.1016/S0921-5093(99)00698-X

Powel, J. (1998). CO2 Laser Cutting. Springer-Verlag London Limited.

Radziejewska, J. (2011). Laserowa modyfikacja warstwy wierzchniej wspomagana nagniataniem [Laser modification of the surface layer assisted by burnishing]. Warszawa: Instytut Podstawowych Problemów Techniki PAN.

Siemiński, P., & Budzik, G. (2015). *Techniki przyrostowe. Druk. Drukarki 3D* [Incremental techniques. Printing. 3D printers]. Rzeszów: Oficyna Wydawnicza Politechniki Rzeszowskiej.

Sobotova, L., & Badida, M. (2017). *Laser marking as environment technology*. Open Engineering. *De Gruyter.*

Šuba, O. (2018). Modelling of a transient-temperature field in plastics during laser cutting. *Materials and Technology, 52*(1), 19-21.

Yongkang, Z., Jinzhong, L., & Kaiyu, L. (2013). Laser Shock Processing of FCC Metals. In Mechanical Properties and Micro-Structural Strenthening Mechanism. Berlin: Springer.

Zawieska, W. M. (2004). Ocena ryzyka zawodowego. 1. Podstawy metodyczne [Risk assessment. 1. Methodological basics]. Warszawa: CIOP PIB.

Ziętek, B. (2009). *Lasery [Lasers]*. Toruń: Wydawnictwo Naukowe Uniwersytetu Mikołaja Kopernika.

Chapter 6
Evaluation of Possibilities to Apply Lasers for Adhesive Bonding

ABSTRACT

The chapter presents the results of experimental research conducted by the author in the context of the possibility of introducing a change from traditional bonding technology to an unconventional one by means of laser surface treatment. The results of the analysis of the surface condition of the materials selected for this monograph were described, both in the so-called delivery condition (after rolling) and after traditional and laser surface treatment. The chapter also provides a comparison of the energy state of the surface to be glued after mechanical, chemical, and laser treatment. The results of bonding were shown in which static strength on the joint wall was taken as a criterion of evaluation. Additionally, unusual possibilities of lasers used to create specific surface roughness and useful in bonding operations were presented.

CONVENTIONAL MATERIALS PREPARATION BEFORE ADHESIVE BONDING

As it was presented in Chapters 3 and 4, adhesive bonding is a multi-stage process in which the initial step is to prepare the part surface before applying the adhesive compound. Depending on the type of the material, a specific treatment is selected, for example:

DOI: 10.4018/978-1-7998-4634-5.ch006

- in the case of steel, these are primarily mechanical treatments (sandblasting), enabling a specific geometric structure of the surface (roughness) to be obtained, and making it possible for the glue to penetrate irregularities (mechanical adhesion) due to excellent conditions,
- in the case of aluminum alloys, these are chemical methods involving the immersion of objects in the solutions of acids, alkali or salts.

The surface preparation stage is a laborious one, as well as inconvenient and dangerous from the point of view of human safety and environmental impact.

Global trends, related to the search for better machining techniques, indicate the need to replace troublesome technologies with others. This chapter presents the effects of experimental work, giving rise to the belief that it is possible to effectively prepare the surface before bonding using a laser beam. Examples of beneficial effects of such processing are visible on steel and aluminum surfaces.

Surface Properties of Rolled Materials

The most convenient choice of materials for adhesive bonding are those with a flat structure, therefore four types of materials were selected to be described in this monograph (according to the US markings) (Figure 1):

- structural steel for quenching and tempering AISI 4130,
- corrosion-resistant steel 430TI,
- aluminum alloy 2017A,
- aluminum alloy 5083,

all in the form of rolled sheets.

They differ in chemical composition and areas of application as shown in Table 1 and 2 (Dobrzański L. A. (2001).

Materials used as semi-finished products for the production of various parts have a certain condition of surface layer. As described in the earlier part of this monograph, it is beneficial that the material surface has cavities (valleys) and elevations that serve as penetration points for the adhesive, especially in

Figure 1. Four types of materials in experimental studies

Table 1. Chemical composition and application of steels used in the experiment

AISI4130 steel		430TI steel	
Chemical composition		Chemical composition	
Element	**Mass concentration, %**	**Element**	**Mass concentration, %**
C	0.22 – 0.29	C	≤ 0.08
Mn	0.40 – 0.70	Mn	≤ 1.00
Si	0.17 – 0.37	Si	≤ 1.00
P	≤ 0.035	P	≤ 0.040
S	≤ 0.035	S	≤ 0.015
Cr	0.80 – 1.10	Cr	16.0 – 18.0
Ni	≤ 0.30	Ni	-
Mo	0.15 – 0.25	Mo	-
V	≤ 0.05	V	-
W	≤ 0.20	W	-
N	-	N	-
Al	-	Co	-
Cu	≤ 0.25	Cu	-
As	≤ 0.08	Ti	-
		Al	-
Application: This steel has very good mechanical properties, is mainly used for welded parts in the aerospace industry, for axles, crankshafts, engine connecting rods, turbine parts.		**Application:** This steel is considered as difficult to weld, its main use is for the production of tanks, cutlery, pipelines, chemical fittings, etc.	

the case of mechanical adhesion. By penetrating the recesses, specific clips are created, thereby improving the strength of the joint.

Examples of smooth surface materials, which do not have properties of mechanical adhesion, are semi-finished steel products in the form of sheets, often used in various branches of the industry for bonded structures. Sheets are available, among other things, after rolling operations which make the

Table 2. Chemical composition and application of aluminum alloys used in the experiment

2024A aluminum alloy		**5083 aluminum allloy**	
Chemical composition		**Chemical composition**	
Element	**Mass concentration, %**	**Element**	**Mass concentration, %**
Si	≤ 0.20	Si	≤ 0.40
Fe	≤ 0.50	Fe	≤ 0.40
Cu	3.5 – 4.9	Cu	≤ 0.10
Mn	0.30 – 0.90	Mn	0.40 – 1.00
Mg	1.20 – 1.80	Mg	4.00 – 4.90
Cr	≤ 0.10	Cr	0.05 – 0.25
Ni	-	Ni	-
Zn	≤ 0.25	Zn	≤ 0.25
V	-	V	-
Ti	≤ 0.15	Ti	≤ 0.15
Al	others	Al	others
Others	≤ 0.15	Others	≤ 0.15
Application: Used in aircraft structures, rivets, small industrial products, wheel rims.		**Application:** Components and structures of rolled products used in the automotive, shipbuilding, aerospace, cryogenic, missile and armor plate industries.	

surface too smooth, therefore rolled parts should be prepared before bonding. AISI 4130 steel and 2024 alloy are matt, whereas 430TI steel and 5083 alloy are glossy. Steel, especially, has a mirror-like surface, which makes it even more difficult to bond such surfaces.

Microscopic observations of rolled sheets show that they do not have a regular cavity and gradient patterns and such issue concerns different types of materials. Presented in Figure1 are the pictures of steel sheet surface produced from AISI 4130 structural steel, 430TI stainless steel, and 2024A and 5083 aluminum alloys. Regardless of their chemical composition, the surface is characterized by slight roughness (Figure 2 and 3).

Figure 2. Appearance of materials after rolling: a) steels, b) aluminum alloys

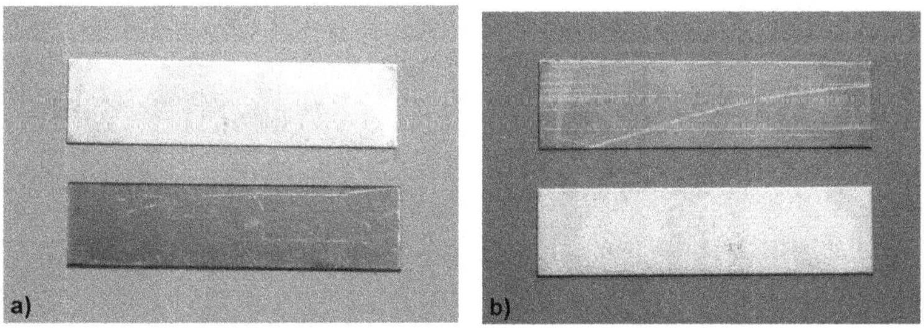

Figure 3. Microscopic image of the surface of rolled materials: a) AISI 4130 steel, b) 430TI steel, c) 2024A alloy, d) 5083 alloy (image magnification 500x)

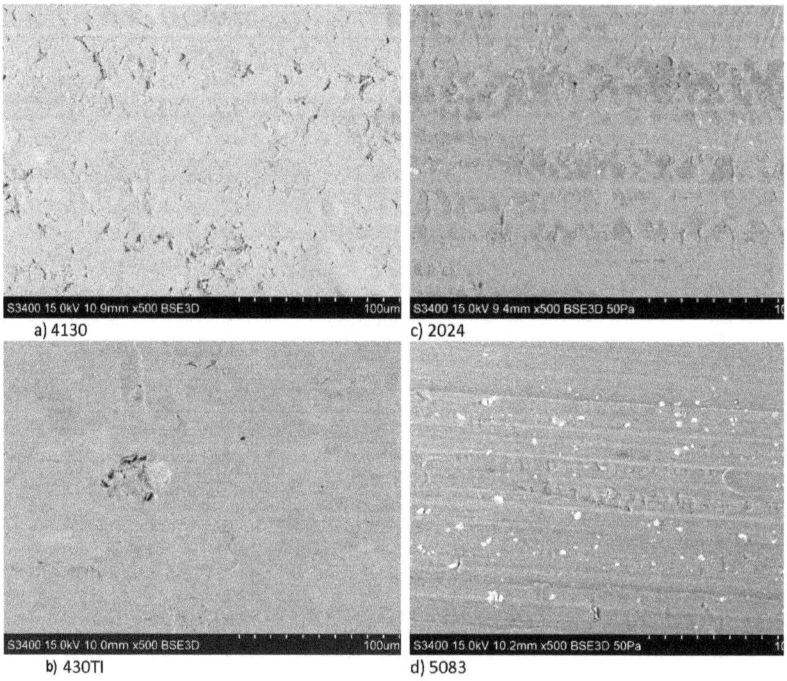

The surface roughness of the alloys varies. For adhesive bonding operations, considering mechanical adhesion, the following parameters were analyzed:

a) amplitude, given in μm:
   Rp - height of the highest elevation of the profile,
   Rv - depth of the lowest profile recess,
   Rt - total height of the profile,
   Ra - mean arithmetic deviation of the profile, mean arithmetic deviation of the profile from the average line, measured along the measuring distance,
b) arrangement
   RSm - mean spacing of local profile elevations, given in mm.

Surface roughness measurement results are shown in Table 3.

Examples of roughness profiles of rolled surfaces are shown in Figure 4.

The energy state of rolled surfaces, important in terms of proper adhesion, can be determined using a goniometer device, e.g. a PG3 goniometer (Figure 5a,

Table 3. Surface roughness of materials after rolling

Steels					
AISI 4130			**430TI**		
Parameter		**Average value**	**Parameter**		**Average value**
Amplitude parameters, μm	**Ra**	0.468	Amplitude parameters, μm	**Ra**	0.042
	Rp	1.880		**Rp**	0.175
	Rv	1.340		**Rv**	0.515
	Rt	5.220		**Rt**	1.670
Arrangement parameter, mm	**RSm**	0.115	Arrangement parameter, mm	**RSm**	0.717
Aluminum alloys					
Aluminum alloy 2024A			**Aluminum alloy 5083**		
Parameter		**Average value**	**Parameter**		**Average value**
Amplitude parameters, μm	**Ra**	0.343	Amplitude parameters, μm	**Ra**	0.233
	Rp	0.960		**Rp**	0.900
	Rv	1.220		**Rv**	0.788
	Rt	2.730		**Rt**	2.380
Arrangement parameter, mm	**RSm**	0.055	Arrangement parameter, mm	**RSm**	0.079

b). The wettability of the surface is determined by measuring the contact angle with two liquids: water and diiodomethane in each case under consideration. According to the state of technical knowledge, it is advantageous when the measuring liquids have a high degree of variation due to polar and dispersion component. On the basis of the results of the experiment during which 4 μl droplets are applied, surface free energy is calculated (Figure 5c, d). The results of such experimental measurements and calculations are given in Table 4.

The graphical comparison of contact angle measured with water and diiodomethane use, SFE, polar and dispersion components are shown in Figure 6.

Surface Condition of Steel After Machining

Due to the unfavorable conditions of adhesion, both specific and mechanical, it is necessary to prepare the steel surface using abrasive methods before bonding. Most often, the surface is prepared manually with abrasive paper or mechanically by sandblasting. When sanded with abrasive paper and sandblasted with 80A corundum, steel surfaces have a roughness greater than

Figure 4. Surface roughness profiles of the selected materials after rolling: a) AISI 4130 steel, b) 430TI steel, c) 2024A aluminum alloy, d) 5083 aluminum alloy

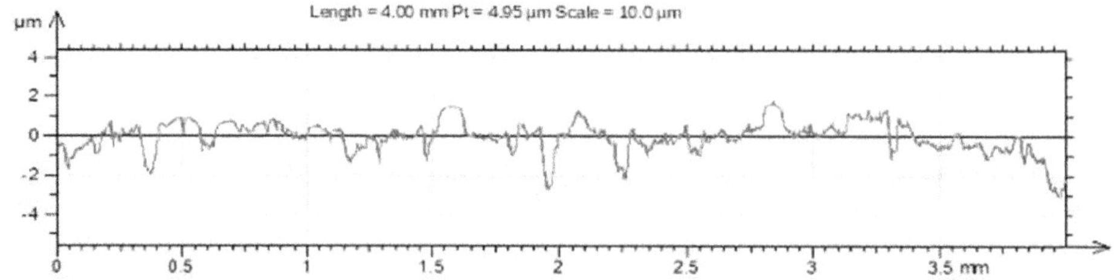

a) AISI 4130 steel

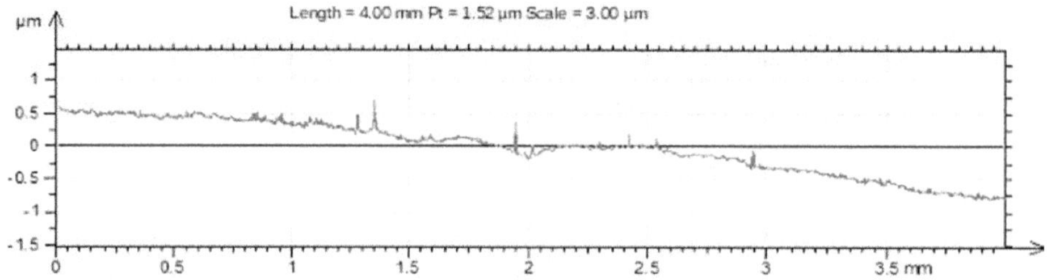

b) steel 430TI

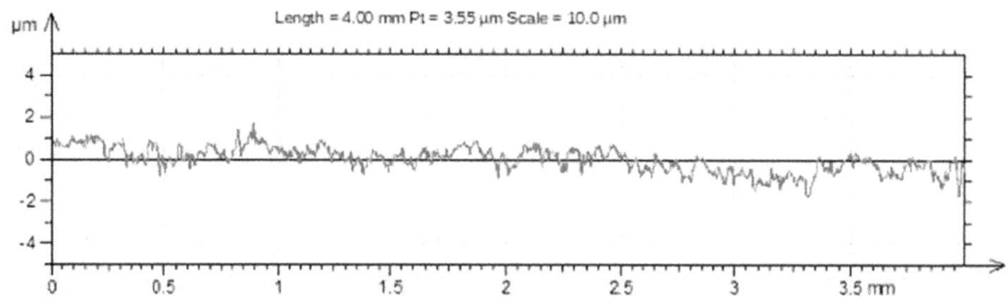

c) 2024A

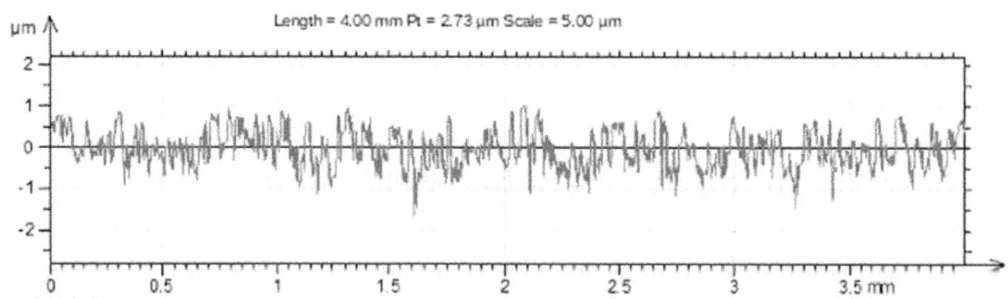

d) 5083

Figure 5. Measurements of contact angle: a) measuring station, b) software window for visualization of measurements, c) specimen with water droplets (white), d) specimen with diiodomethane droplets (yellow)

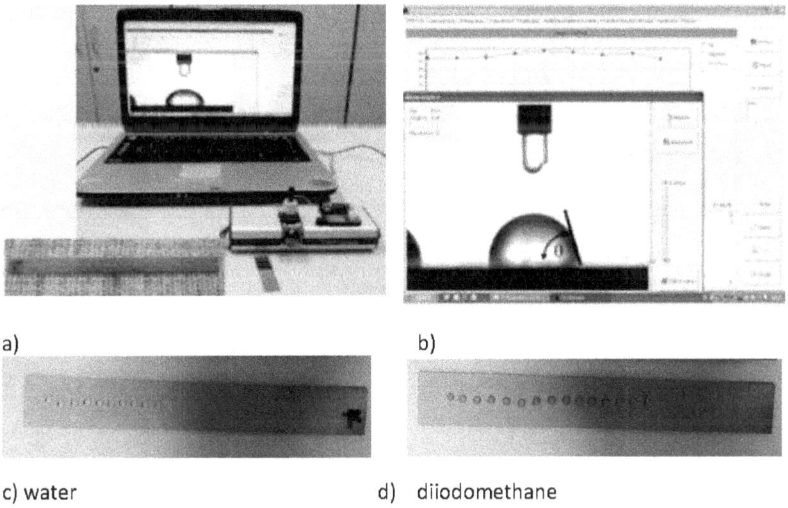

Table 4. Wettability and SFE of rolled material surfaces

Material	Θw, °	Θd, °	γp, mJ/m²	γdis, mJ/m²	SFE, mJ/m²
AISI 4130	80,6	45,9	8,7	36,4	46,0
430TI	85,0	39,9	6,4	40,2	46,6
2024A	83,9	48,1	5,4	42,6	43,8
5083	86,6	34,5	5,4	42,6	48,0

Figure 6. The presentation of the measurement results of contact angles (left), SFE and its components on rolled materials (right), where: a) 4130 AISI steel, b) 430TI steel, c) 2024A aluminum alloy, d) 5083 aluminum alloy

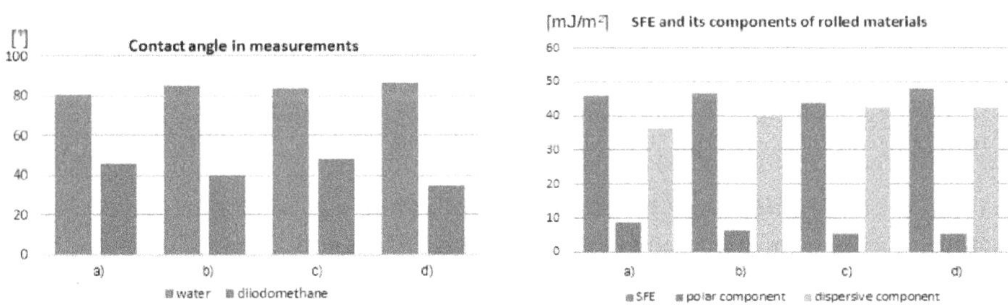

in the delivery condition (after rolling). This is advantageous due a wider surface area to which the adhesive will be in contact in the bond.

The surface appearance and results of surface roughness measurements of steel after sandblasting are shown in Figure 7 and 8, and Table 5.

Figure 7. Microscopic image of the surface after sandblasting: a) AISI 4130 steel, b) 430TI steel (image magnification 500x)

a) b)

Table 5. Surface roughness of steel after sandblasting

Steels					
AISI 4130			**430TI**		
Parameter		**Average value**	**Parameter**		**Average value**
Amplitude parameters, μm	**Ra**	3.01	Amplitude parameters, μm	**Ra**	1.96
	Rp	8.02		**Rp**	6.10
	Rv	9.43		**Rv**	8.41
	Rt	22.3		**Rt**	20.1
Arrangement parameter, mm	**RSm**	0.098	Arrangement parameter, mm	**RSm**	0.085

Surface roughness increases after blasting and sandblasting and, what is more, the structure of elevations and cavities becomes more regular, although it is a coincidence due to the stochastic nature of loose abrasive blasting.

Abrasive treatment of the surface changes its wettability. SFE of steel AISI 4130 increases after sandblasting, while in the case of steel 430 TI slightly decrease (Table 6).

Figure 8. Roughness profiles of steel surface after abrasive machining (sand blasting): a) AISI4130, b) 430 TI

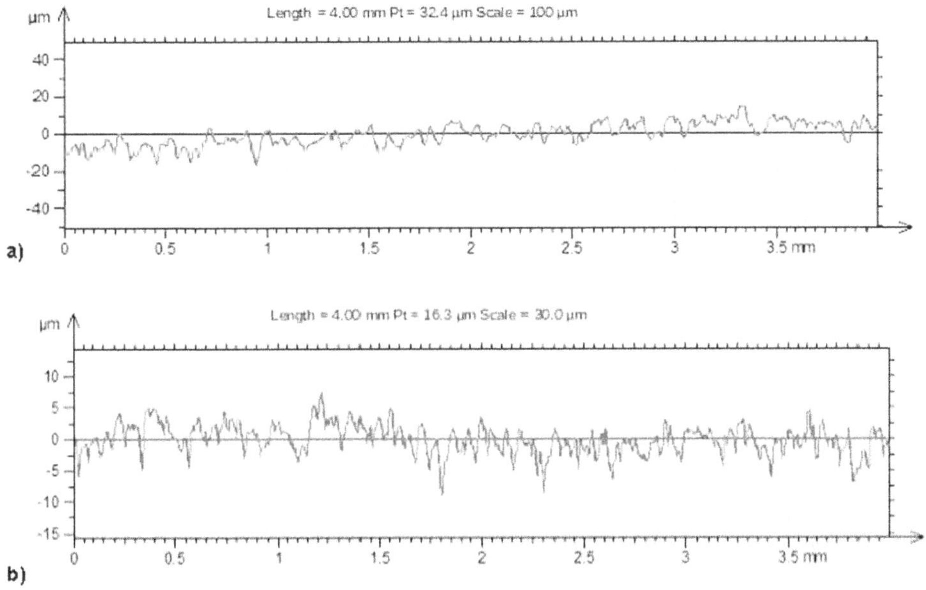

Table 6. The results of measurements of steel surface wettability after sandblasting

Material	Θw, °	Θd, °	γp, mJ/m²	γdis, mJ/m²	SFE, mJ/m²
AISI 4130	112.7	41.4	0.9	39.1	40.0
430 TI	134.3	46.2	5.8	36.4	42.2

Comparative illustrations of the obtained results are shown in Figure 9.

Comparing the results of contact angle measurements and the calculated surface energy, it can be concluded that after sandblasting:

a) for AISI4130 steel:
contact angle – measurement by water increased from 80.6 to 112.7°,
contact angle – measurement by diiodomethane decreased from 45.9 to 41.1°,
SFE decreased from 46.0 to 40.0 mJ/m^2,
polar component decreased from 8.7 to 0.9 mJ/m^2,
dispersive component increased from 36.4 to 39.1 mJ/m^2,

Figure 9. Contact angle (a), SFE and its components (b) after sandblasting

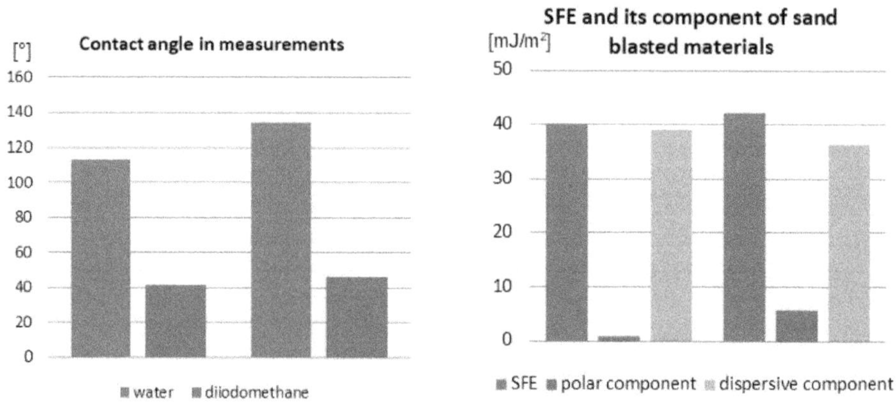

a) *b)*

b) for 430TI steel:
contact angle – measurement by water increased from 85.0 to 134.3°,
contact angle – measurement by diiodomethane increased from 39.9 to 46.2°,
SFE decreased from 46.6 to 42.2 mJ/m^2,
polar component decreased from 6.4 to 5.8 mJ/m^2,
dispersive component decreased from 40.2 to 36.4 mJ/m^2.

The process of sandblasting of the analyzed steels have resulted in the development of the surface (a wider surface area) and the greater roughness which may improve mechanical adhesion. After sandblasting, the total height of roughness (profile Rt) increased:

- for AISI4130 steel from 5.22 to 22.3 μm,
- for 430 TI steel from 1.67 to 20.1 μm.

However, despite the treatment which is considered to be effective for surface preparation before bonding, specific adhesion conditions have deteriorated. Surface free energy has decreased and in most cases, contact angles tended to be of poorer wettability (more than 90°).

Surface Condition of Aluminum Alloys After Chemical Processing

Chemical treatment is primarily applicable in the surface preparation of aluminum alloys. As a result, a favorable oxide layer is formed on the surface of the material, which is firmly bound to the substrate. Oxides have a porous structure, which improves the adhesive properties of the material. The structure of the surface layer after chemical treatment of 2024A and 5083 alloys is shown in Figure 10.

Figure 10. Pictures of the surface of chemically treated aluminum alloys: a) 2024A, b) 5083 (image magnification 500x)

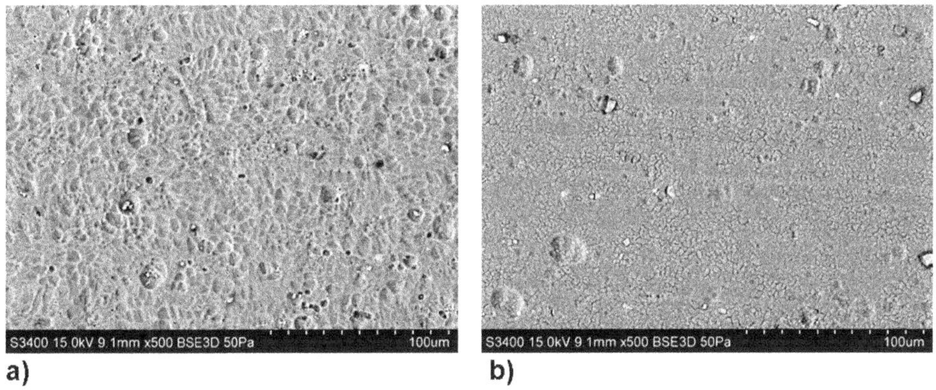

Due to the formation of a layer of a different structure than the material core, a change in surface roughness (Table 7 and Figure 11) and energy state of alloys is also observed.

As in the case of steel abrasive processing, the consequence of the chemical treatment of aluminum alloys is a change in energy state.

The results of measurements of the contact angle with two liquids and the SFE are given in Table 8.

Comparing the results of contact angle measurements and the calculated surface energy, it can be concluded that after chemical treatment:

Table 7. Surface roughness of chemically treated aluminum alloys

Aluminum alloys					
Aluminum alloy 2024A			Aluminum alloy 5083		
Parameter		Average value	Parameter		Average value
Amplitude parameters, µm	Ra	0.337	Amplitude parameters, µm	Ra	0.206
	Rp	1.165		Rp	0.918
	Rv	1.370		Rv	1.207
	Rt	3.803		Rt	3.037
Arrangement parameter, mm	RSm	0.051	Arrangement parameter, mm	RSm	0.065

Figure 11. Roughness profiles of surfaces of aluminum alloys after chemical treatment: a) 2024A, b) 5083

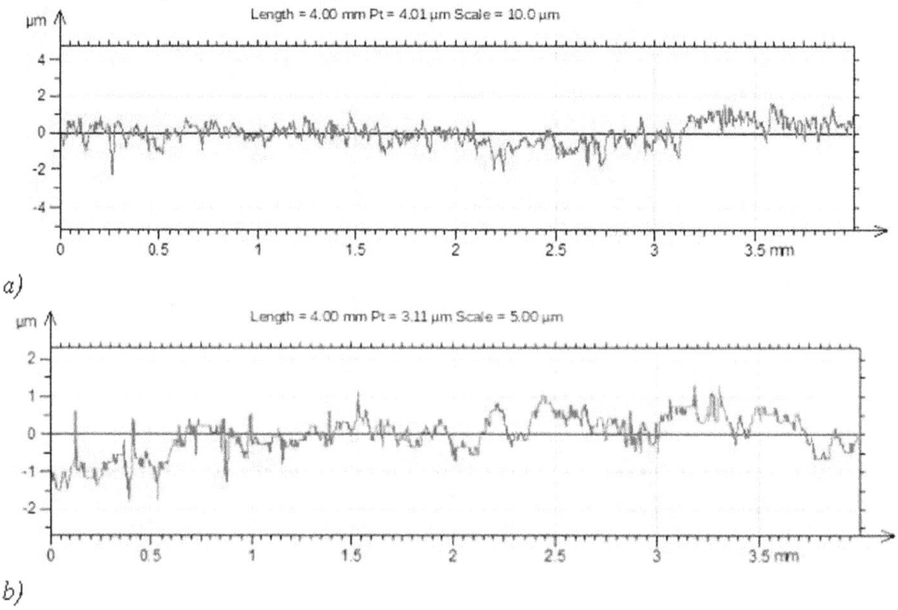

a)
b)

Table 8. Results of measurements of wettability of aluminum alloys after alodinating

Material	Θw, °	Θd, °	γp, mJ/m²	γdis, mJ/m²	SFE, mJ/m²
2024A	63.7	46.0	16.6	37.3	53.9
5083	38.2	61.3	32.6	29.8	62.5

Figure 12. Contact angle (a), SFE and its components (b) after chemical treatment

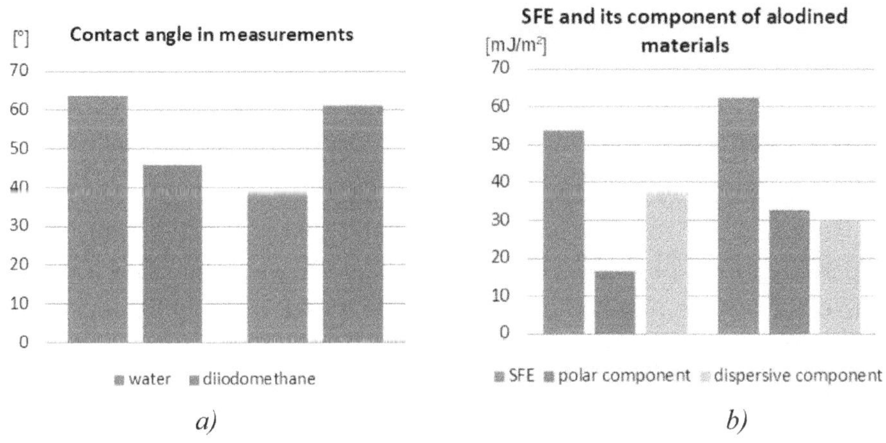

a) for 2024A aluminum alloy:
   contact angle – measurement by water decreased from 83.9 to 63.7°,
   contact angle – measurement by diiodomethane decreased from 48.1 to 46.0°,
   SFE increased from 43.8 to 53.9 mJ/m^2,
   polar component increased from 5.4 to 16.6 mJ/m^2,
   dispersive component decreased from 42.6 to 37.3 mJ/m^2,
b) for 5083 aluminum alloy:
   contact angle – measurement by water decreased from 86.6 to 38.2°,
   contact angle – measurement by diiodomethane decreased from 34.5 to 61.3°,
   SFE increased from 48.0 to 62.5 mJ/m^2,
   polar component increased from 5.4 to 32.6 mJ/m^2,
   dispersive component decreased from 42.6 to 29.8 mJ/m^2.

The process of chemical treatment of the analyzed aluminum alloys have resulted in the development of the surface (a wider surface area) and the greater roughness which may improve mechanical adhesion. After alodining, the total height of roughness (profile Rt) increased:

- for 2024A aluminum alloy from 2.730 to 3.803 μm,
- for 5083 aluminum alloy from 2.380 to 3.037 μm.

The result of chemical treatment is the favorable energy state of the surface (SFE increased), while the mechanical adhesion conditions have changed slightly.

STATIC SHEAR STRENGTH OF ADHESIVE BONDS AFTER CONVENTIONAL PROCESSING

In experimental studies, static shear strength is often used aimed at assessing the efficiency of the preparatory processing and determining the strength of the adhesive bond. The surfaces prepared for bonding by means of mechanical or chemical method with a chosen adhesive, form a single-lap joint (as in the example described in this monograph based on experimental studies).

The bonded elements are cut from sheets, rolled of 1 mm thickness, as in the case in question, with dimensions of 100x25 mm.

The joints prepared in such a way are broken off on the testing machine (Figure 13).

Figure 13. Static shear strength tests: a) adhesive bonding carried out for tests, b) Zwick Roel testing machine

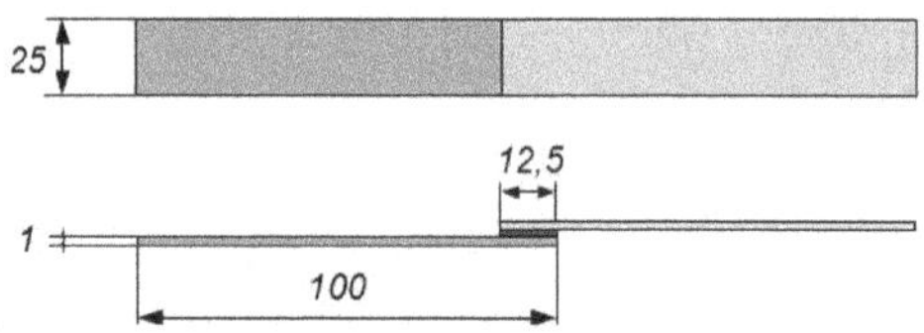

a)

b)

In addition to the information about the force needed to break the connection, this test allows to determine the nature of the break (either adhesive or cohesive - Figure 14), and to determine the strength of the joint, by dividing the average breaking force of Ptmed by the field of the A_0. In the presented examples, the overlap had the dimension of 12.5 x 25 mm, which gives $A_0 = 312.5\ mm^2$.

Figure 14. The examples of broken joints and evaluation of the break characteristics: a) aluminum alloy 2024 A after chemical treatment - adhesive break, b) AISI steel 4130 - cohesion break

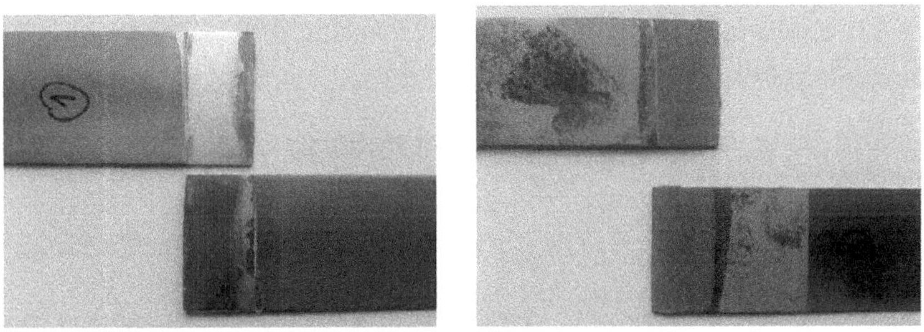

a) b)

The static shear strength may be determined based on a test of at least 3 to 5 samples. The mean value is then calculated and the above dependence is used.

Examples of static shear strength of the joints made with differently prepared surfaces are shown in Table 9 and Figure 15.

STATE OF THE SURFACE AFTER LASER PROCESSING

The application of ablation, evaporation or laser etching phenomena makes it possible to clean surfaces of different types of materials. As a result of the impulsc impact, thc cncrgy is absorbcd, thc surfacc laycr is mcltcd and thc material from the melted zone is rapidly evaporated and ejected in the form of plasma as fragments of material particles and reaction products (Dobrzański L. A., 2001; Burakowski T. & Others., 2006). In this way, various organic and inorganic layers can be removed from concrete, metal, and lacquered objects. This method is applied in the renovation of works of art, where sometimes

Table 9. The results of static strength tests

Value	Sand blasting		Alodining	
	AISI4130	430TI	2024A	5083
Breaking load, N	5365.63	7584.38	2012.5	1720.0
Static strength of the join, N/mm^2	17.17	24.27	6.14	5.51
Commentary: One adhesive was used to prepare single-lap joints (Araldite 2011) The strength of the joint was determined from the dependency: strength = breaking load Ptmed / overlap area A_0 (312.5 mm^2)				

Figure 15. Summary of the results of static shear strength tests of adhesive bonds made with the same adhesive: a) 4130 after sandblasting, b) 430TI after sandblasting, c) 2024A after alodining, d) 5083 after alodining

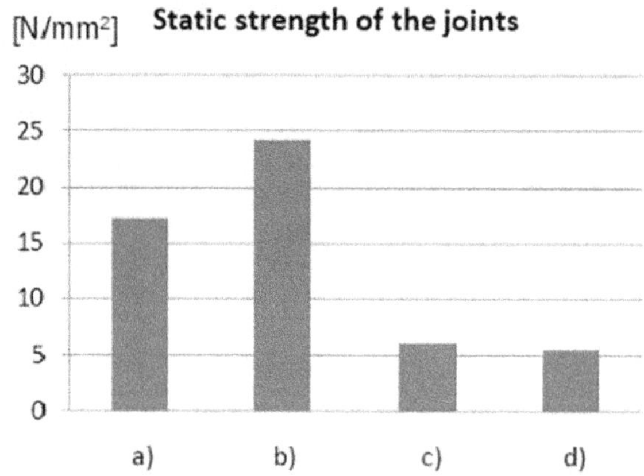

centuries-old multi-layers are successfully removed without damaging the surface (Ciecińska B., 2016; Marczak J., 2004; Major B., 2002). In addition, such method is successfully applied in adhesive bonding technology, where it is possible to degrease and clean the surface before applying the adhesive.

Laser surface preparation requires that the beam is evenly distributed over the surface. Due to the nature of the work, it is relatively easy to control the movement of the impulse laser, in which it is possible to modify:

- laser power, W,
- pulse frequency, Hz,
- scanning speed, mm/s.

When a robotized workstation is not provided, a conventional machine tool can be used for surface treatment, in which a laser head fixed in place of the tools will not rotate (e.g. a vertical milling machine with off-rotation), yet the worktable will provide motion at the set speed.

An example of such a technological solution is shown in Figure 16. The optical fiber laser beam was mounted in a special holder and directed vertically downwards, the workpiece lay on the table provided for movability with respect to the beam.

Figure 16. Arrangement of laser processing: a) optical fiber laser head, b) surface treatment

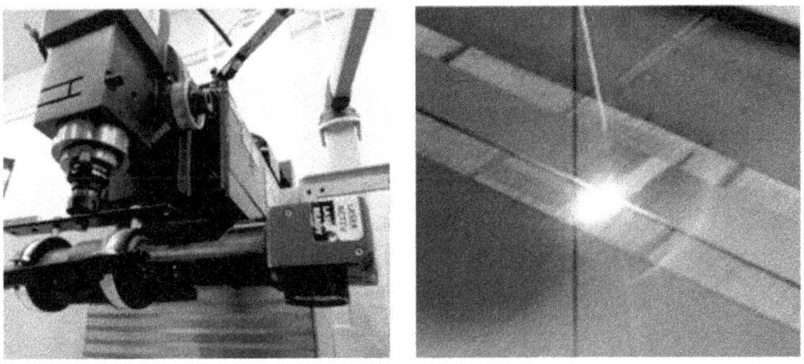

a) b)

Using the pulsed fiber laser may not only result in a clean surface (accumulation of oil, dirt - was evaporated), but at the same time, it is possible to shape it according to preferences.

The heads of different power are used, and the processing effects are varied (Figure 17).

Laser cleaning allows to resign from the inconvenient chemical treatment of aluminum alloys. For the given experimental test variants, surfaces with relatively small roughness are obtained, yet with favorable energy status (Table 10).

The change of laser parameters affected the free surface energy (SFE). However, depending on the device, it is possible to select a range of optimal parameters to obtain the highest surface energy value.

In most cases, the optimal energy state of the surface is not related to the smallest roughness Rt. In general, SFE had the highest value for the lowest pulse frequency.

Figure 17. The examples of surface creation after laser beam impact: a) 2024A – the same laser, different scanning speed and impulse frequency: 100 mm/s, 50kHz on the left; 1000 mm/s 200kHz on the right; b) 430TI – the same laser and impulse frequency, the speed of scanning different: 1000 mm/s on the left, 8000 mm/s on the right

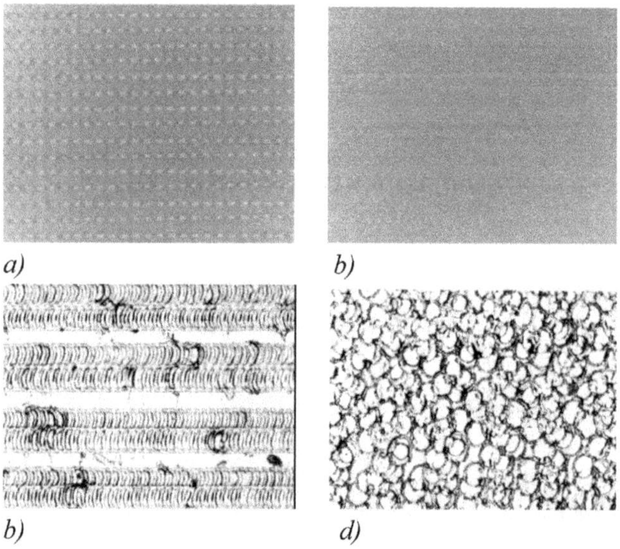

Table 10. Example of experimental results for a selected aluminum alloy – 2024A

Rt	Impuls frequency kHz	Speed of scanning mm/s	Θw °	Θd °	SFE mJ/m²	γdis mJ/m²	γpol mJ/m²
3.21	10		10.1	16.8	**84.6**	48.7	35.9
3.51	15	1000	12.5	15.4	84.4	49.0	35.4
2.42	20		24.0	38.0	75.3	43.6	31.7
3.16	10		36.8	8.1	77.1	50.3	26.8
3.66	15	2000	13.1	13.3	**84.6**	49.5	35.1
3.12	20		7.9	33.8	80.9	42.9	38.0
3.12	10		13.6	13.8	**84.5**	49.4	35.1
3.42	15	4000	17.1	20.9	82.3	47.6	34.7
3.52	20		19.5	37.1	77.4	41.5	35.9
3.56	10		7.9	10.4	**85.8**	50.0	35.0
4.14	15	6000	30.0	34.8	74.2	42.5	31.7
3.09	20		16.6	38.9	77.7	40.7	37.0

Θw min = 7.9°
Θd min = 8.1°
SFE max = 85.8 mJ/m²

The laser beam can be used as an effective tool to change energy state.

The surface roughness is not always significantly reduced. However, in comparison with small surface roughness after rolling, the surface after laser cleaning has a more favorable surface energy and is better

STRENGTH TESTS

Table 11. Comparison of SFE best results after different treatments (given in mJ/m²)

Material	Rolled	Sand blasted or chemical treated	Laser treated
AISI 4130	46.0	40.0	63.3
430 TI	46.6	42.2	76.0
2024A	43.8	52.4	85.8
5083	48.0	40.7	75.5

The surface structure obtained after laser treatment is specific yet beneficial in terms of roughness and energy state. This condition translates into static shear strength of the connections in which the surfaces were treated with a laser.

Preparation of the surface with a laser, following the above arrangement, allows to improve the energy status of the surface. The overlapping joints have also satisfactory static shear strength than after mechanical or chemical surface preparation.

Examples of connection strength values are given in the Table 12 (for 430TI steel as an example).

Experimental studies show that the use of laser for surface treatment can produce beneficial results. Although there is little difference between the static shear strength figures, the benefits of using lasers are significant, namely shorter preparation time before bonding and lack of negative impact on the environment.

Special Case of Mechanical Adhesion Formation

The use of a pulse fiber optic laser head makes it possible to make deeper marks on the metal surface. The following examples show traces of a double transition (I - along with the sample, II - across) that forms a specific grid.

Table 12. The results of static strength test on laser treated samples

Impuls frequency kHz	Speed of scanning mm/s	Breaking load, N	Static strength of the join, N/mm²
10 15 20	1000	4789.66 4742.32 5216.54	15.34 15.17 **16.70**
10 15 20	2000	4003.83 5037.26 4819.52	12.81 **16.12** 15.42
10 15 20	4000	4964.10 4486.24 5255.52	15.88 14.36 **15.98**
10 15 20	6000	5038.18 4742.32 5255.52	16.12 13.35 **16.82**
Commentary: One adhesive was used to prepare single-lap joints (Araldite 2011) The strength of the joint was determined from the dependency: strength = breaking load / overlap area (312.5 mm²)			

The use of a conventional machine tool table drive creates a controllable variation in dimension between the individual lines (Figure 18 and Table 13).

This unconventional method of surface preparation allows for the creation of regular grooves for the needs of mechanical adhesion. By controlling the speed of moving the sample with respect to the laser beam, it is possible to obtain specific distances between the successive lines, as well as the depth of the grooves (due to the variable time of the energy beam on the surface). The example of AISI 4130 steel, presented in Table 14, shows the result of surface roughness measurements, the regular arrangement of cavities and evelations and the depth of the grooves.

As can be seen, with the increase of the sample speed movement, in relation to the head and the laser beam, the distance between the grooves increases, whereas their depth slightly decreases. This is logically related to the time that influences laser energy on the material - the shorter the time, the shallower the place of influence. Creating the surface to be bonded in this way can yield beneficial results in terms of improved static shear strength of the joints. Examples of experimental results are given in Table 15, where specimens from 430TI steel and 2024A and 5083 aluminum alloys were presented using the same method.

Surface number 1 can be classified as the optimal example of this method, where the deepest grooves were located closest in relation to each other. The

Figure 18. Samples made for unconventional surface structure before adhesive bonding: a) diagram of surface scanning method, b) four options of experiment, c) an example of special grid

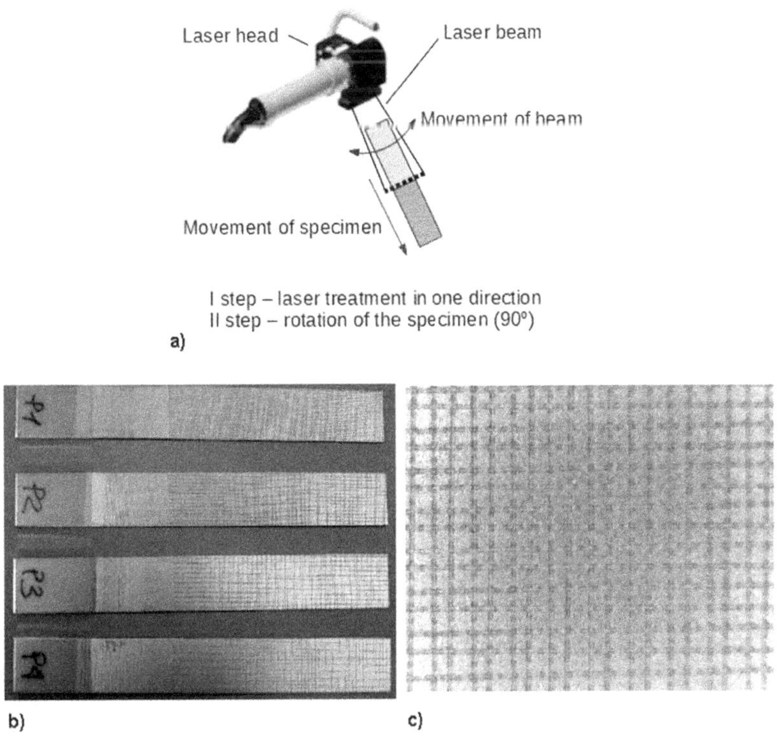

Table 13. Variants of laser processing of surfaces from the figure above

No.	Laser operating parameters	Machine table feed, mm/min.
1	Scanning width - 50 mm Laser power 100% - 50W Pulse frequency - 50 kHz Scan speed - 150 mm/sec	115
2		175
3		250
4		375

results of the experiment confirm the hypothesis of mechanical adhesion. The surfaces of the samples were not degreased earlier, thus the side surfaces of the grooves - clean from the dirt after beam penetration - and their depth were sufficient to obtain satisfactory bond strength.

Figure 19. The surface profiles of number: a) 1, b) 2, c) 3, d) 4 of samples from Table 13

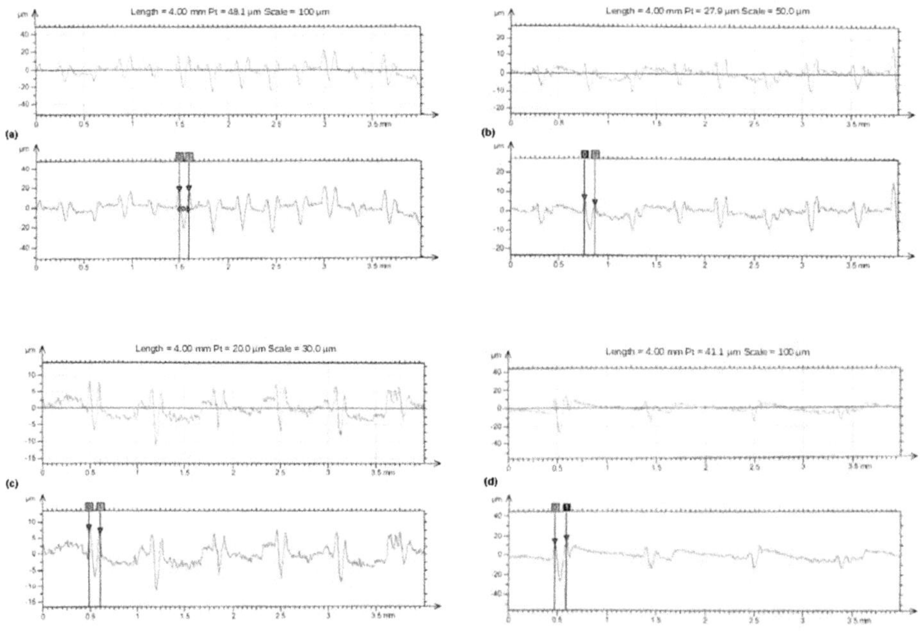

Table 14. Distances between the grooves and size of the cavity (mean values from 3 measurements)

No. of specimen	**1**	**2**	**3**	**4**
Distance, mm	0.102	0.109	0.116	0.122
Depth, µm	1.99	1.73	1.15	0.99

*Table 15. Results of the static shear tests **

Material	**Number of specimen**			
	1	**2**	**3**	**4**
AISI 4130 steel	**18.760**	17.460	17.850	16.785
430 TI steel	**24.678**	21.927	19.280	20.327
2024A aluminum alloy	**25.182**	20.604	20.295	16.757
5083 aluminum alloy	**16.373**	14.200	13.078	12.397

* for the purposes of the experiment, single-lap joints were made, with an overlap area of 312.5 mm^2, with the same adhesive as in the other cases described in Chapter 6 (the numerical value given in the table is the arithmetic mean of 5 measurements)

DURABILITY OF THE LASER EFFECT: SURFACE FREE ENERGY VARIATIONS

The effects of surface treatment of coating materials, including adhesives, are not always stable over time. An example of a changing parameter is the surface free energy (SFE). Unfortunately, it may decrease over time, thus worsening the adhesion conditions. This is essential for the planning and implementation of manufacturing processes. They often require the positioning of objects, precise folding or transport. In the case of a specially generated energy state, such products should be designed for such operations with a short lead time.

An example of a change in SFE is previously tested steel 430TI. Experimental studies show that in the case of SFE, the value of SFE falls by about 20% after 1 hour after the end of machining.

To determine if this negative effect applies only to laser samples, it is worth comparing the results with non-treated, mechanically and laser machined surfaces, which are shown in Table 16.

Table 16. Values of SFE of 430TI steel – the change in time

Option	The measurement taken shortly after processing				
	θw	θd	SFE	γdis	γpol
Rolled	60.4	36.9	**58.8**	41.6	17.3
Sand blasted	67.5	14.1	**60.1**	49.3	12.5
L1	26.5	30.6	**76.8**	44.2	32.6
L2	27.7	28.5	**76.9**	45.0	31.9
L3	49.5	32.4	**65.9**	43.5	22.5
Option	**Measurement taken 1 hour after processing**				
	θw	**θd**	**SFE**	**γdis**	**γpol**
Rolled	55.8	35.4	**55.8**	42.2	19.4
Sand blasted	67.5	14.1	**61.8**	49.3	12.5
L1	60.6	38.0	**58.3**	41.1	17.3
L2	60.6	37.3	**58.6**	41.4	17.2
L3	73.5	41.9	**50.7**	39.3	11.4

L1 – laser treatment with the speed of the specimen – 35 mm/min.
L2 – as above, 115 mm/min.
L3 – as above, 179 mm/min.

The results show that the problem of energy state instability concerns all machining methods (only in the delivery status - the energy increases slightly after rolling). The decrease in SFE and its components is relatively small for conventional machining, yet fairly significant for laser machining. This is an important signal in the context of technological change. The elimination of the so-called wet chemistry or inconvenient mechanical processing entails consequences not only in the form of benefits for human health and environmental safety, but also in the form of the need to reorganize other production processes. However, the concept of industrial development described in Chapter 8 gives rise to hope that the situation, in which the energy state of the surface before adhesive bonding deteriorates, will not cause an obstruction to the implementation of flexible and effective technological processes.

CONCLUSION

The analysis carried out in the field of adhesives shows that it is possible to change the previously applied methods of preparing the surface to be glued together for laser treatment. The selected steels, so far typically machined, and aluminum alloys prepared chemically after laser treatment were characterized by a similar or even better energy state of the surface and wettability. Static shear strength tests also gave positive results. Connections made according to the method described in this chapter had satisfactory properties.

The presented possibility of using lasers in adhesive technology may generate measurable benefits. It is possible not only to remove layers and impurities, but also to improve wettability and obtain durable joints. An additional benefit is the elimination of chemical substances, devices, and equipment used for preparatory works and the negative impact on human health and the environment.

REFERENCES

Burakowski, T., Marczak, J., & Napadłek, W. (2006). *Istota ablacyjnego czyszczenia laserowego materiałów* [The essence of ablative laser cleaning of materials]. Warszawa: Prace Instytutu Elektrotechniki WAT, z. 228.

Ciecińska, B. (2016). Laser beam surface pretreatment for the adhesive bond strength improvement. *Strength of Materials*, *48*(4), 561–565. doi:10.100711223-016-9798-4

Dobrzański, L. A. (2001). Zasady doboru materiałów inżynierskich z kartami charakterystyk [Rules for the selection of engineering materials with safety data sheets]. Gliwice: Wydawnictwo Politechniki Śląskiej.

Major, B. (2002). Ablacja i osadzanie laserem impulsowym [Pulse laser ablation and deposition]. Kraków: Wydawnictwo Naukowe AKAPIT.

Marczak, J. (2004). *Analiza i usuwanie nawarstwień obcych z różnych materiałów metodą ablacji laserowej [Analysis and removal of foreign layers from various materials by laser ablation]*. Warszawa: WAT.

Chapter 7
Environmental and Human Aspects of Adhesive Bonding and Laser Treatment

ABSTRACT

This chapter presents the problems of ensuring safe working conditions in modern production plants. It discusses the modern concept of occupational risk and environmental aspects. Considering problems such as human health and the impact of industry on the environment, the analysis of conventional techniques used in adhesive bonding and their proposed alternative in the form of laser processing were presented. Health hazards and environmental nuisance of both described manufacturing techniques were also characterized in this chapter.

MODERN PRODUCTION MANAGEMENT: SAFETY AT WORK AND ENVIRONMENTAL CIRCUMSTANCES

Production is the process of transforming numerous factors into products intended for use and into the derivative results of the initiated processes. It is a specific system of components and relations between them, and relations of transforming input factors into output ones. It is typically a deliberately designed and organized material, energy and information system, exploited by humans to meet specific needs.

DOI: 10.4018/978-1-7998-4634-5.ch007

At the input stage of the system there are factors necessary to start the production process. The so-called input vector consists of: materials, tools, technological devices, energy, technological and organizational knowledge and people. Finished goods or services are expected to leave the system, but there are also by-products of the initiated processes. This may be waste, products not in line with orders or incorrectly executed additional knowledge (Groover M., 2008).

At the output of the system there are finished goods or services as well as by-products of the initiated processes such as waste, products manufactured not in line with orders, faulty products, gained knowledge. In order to receive output products from the input elements, the process of transforming the inputs into outputs must be realized. It is called the production process, and it consists of (Groover M., 2008):

- technological operations,
- transport,
- supervision,
- storage,
- maintenance operations.

Relationships between the inputs and the outputs and the production process play an important role. They enable an effective flow of materials and semi-finished products from warehouses to machining stations as well as the implementation of the technological process and the flow of parts throughout the production facility. These relationships should:

- ensure that all elements of the system work in time,
- ensure the supply of materials and tools,
- reduce losses to a minimum,
- ensure the supply of energy, water, gases, etc.

The functioning of the system, consisting of many inputs, outputs and production processes, is controlled by a person, guided by a certain way of thinking, a certain philosophy, which allows distinguishing the criteria of selecting factors for the production goal (Durlik, 2007; Groover M., 2008).

Different ways of strategic thinking, e.g., based on the concept of total quality management (TQM), are visible in company management. Total management assumes the pursuit of sustainable economic success by companies through

Figure 1. Scheme of manufacturing model

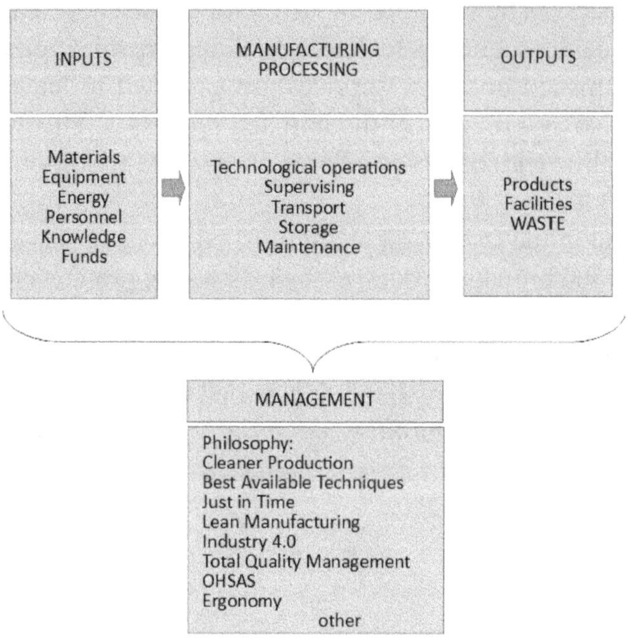

the use of advanced management techniques and tools and the improvement of all activities and processes aimed at satisfying the expectations of external and internal customers. It follows from the definition of TQM that a company (implementing TQM principles) should take into account not only its financial profits and the satisfaction of customers who buy services or goods, but also employee satisfaction (Lewandowski J., 1998).

The basic condition for achieving employee satisfaction is to ensure **safe and healthy living and working conditions**. Today's companies are expected to **manage their health and safety** systems.

Attention is also drawn to the environmental context as a result of using natural factors of productions such as metal ores, gases, oil and transforming them into energy or into production materials. The impact of waste on the natural environment is an extremely important issue, especially in recent years. Waste not only degrades the aesthetic qualities of the landscape but also poses a threat to human life and health by lowering the quality of life (Editorial, 2012).

Environmental management, limiting the negative impact of technologies, substances, both processed, transported and stored, and then utilized or landfilled, is the second, important premise of production management (Geng H, 2004).

Occupational Safety and Risk

More than 80% of accidents in the processing industry are related to the worksite operations. Hazards arise from the construction of machinery, materials and processes, control and operation methods.

Therefore, the issue of ensuring safety requires an analysis of the activities performed in the workplace as well as the material and non-material factors linked to the workplace.

The risk analysis should generally include (Kobyłko G., 2000; Podgórski D., 2001):

- understanding how the equipment works,
- specification of the additional tooling at the workplace,
- identification of the chemical substances necessary to carry out the technological process using a given device,
- identification of environmental factors such as temperature, air quality and radiation,
- risk estimation and assessment,
- the risk reduction strategy for the workplace,
- specification of the collective protection measures used,
- specification of personal protective equipment.

Risk and Hazard: Basic Concepts

Risk analysis should be based on the identification of events that may occur in the workplace and may endanger human safety.

- **Access time** - the time during which an employee is exposed to harmful or onerous factors causing a hazard.
- **Frequency of access** - the number of events in a time unit at which access to the station is expected or required.
- **Defect** - (fault) - condition of an object - tooling, instrumentation, machine - inability to perform the required functions (except for time spent on maintenance and repair or downtime resulting from lack of initiated manufacturing activities).

Among the many defects, there can be fail-to-danger faults - any defects in the machine or its power supply that create a hazardous situation.

- **Information for use** - security measures consisting of communication elements such as text, logos, symbols or diagrams, used separately or together to provide information to the user.
- **Rationally foreseeable misuse** - the use of a product or service in the conditions or for purposes that are not stated by the supplier but that may occur. It is caused by a product, process or service, in combination with other human behavior or as a result of common, but not always correct, practices.
- **Risk** - a combination of the probability of a loss event and its severity.
- **Residual risk** - risk remaining after protective measures have been taken.
- **Tolerable risk** - a risk that has been accepted in given circumstances.
- **Danger zone** - any area in which a person is exposed to the danger of injury or deterioration of health.
- **Hazardous situation** - any situation in which a person is exposed to one or more threats.
- **Damage** (harm) - physical injury or damage to human health, damage to property or degradation of the natural environment.
- **Safety measures** - a combination of risk reduction strategies used to achieve at least tolerable risk.
- **Hazard** - the source of possible injury or deterioration of health.
- **Harmful event** – an occurrence in which a hazardous situation transforms into harm.

Modern Risk Concept

The modern concept of creating a safe working environment is based on the rationalization of occupational risk. Occupational risk is the only parameter that makes it possible to quantify the level of safety. Risk assessment is also the only tool for effective occupational safety management in companies (Badida M. & Others, 2014).

Occupational risk is defined as the possibility of incurring losses in the work process, i.e., the probability of a harmful event or threatening event causing specific damage. The probability of an event occurring is related to the simultaneous occurrence of a specific dysfunction in the workplace structure and the exposure of people or objects to that dysfunction.

With reference to the production machinery in use, there is a predominant risk of injuries related to the machinery elements that are:

- in motion,
- under voltage,
- under considerable pressure,
- hot or cold,
- sharp, rough,
- damaged or operating in a way that is not compatible with the intended use (unpredictability of the mechanisms, incorrect operating ranges, ctc.).

In the case of chemical substances, the health risks arising from the use of, e.g., packaging with liquids, gases or loose products, possible reactions with the environment, emissions to the environment, etc., are particularly important.

Health risks are difficult to monitor, but should be well recognized, especially if they arise from the particular nature of the job. Despite different ways of protecting people's health and lives, there are still residual risks. Work safety management is about consciously taking or rejecting residual risks. Residual risks should remain at a level reduced to a deliberately acceptable and reasonable minimum under the circumstances (in view of the current state of knowledge and technology, costs, production organization, education and skills, staff culture and awareness).

Tolerable risk is considered to be a basic level below which one cannot go.

Manufacturers are responsible for the level of hazards generated by machines and the hazards arising from their operation.

At present, the essence of any action to reduce risk is accident prevention and health protection. Therefore, the starting point for safety management should be the recognition and understanding of production processes and the likelihood of accidents and occupational diseases.

The modern approach to promoting respect for the human being is expressed in the provisions of law and international law, in the context of the globalization of lifc, production and consumption. An example is the international standard OHSAS 45001 (Occupational Health and Safety Management Systems). According to the guidelines of the standard, a company may obtain confirmation from an independent entity that it has implemented the so-called security policy in the production plant. The safety policy shall include the commitment of the management to:

- prevention of accidents at work, occupational diseases and potential accidents and incidents,
- striving for continuous improvement in occupational health and safety,
- compliance with legal requirements (some recommendations are mandatory in some countries and voluntary in others),
- continuous improvement of activities in the field of occupational health and safety,
- ensuring adequate resources and measures,
- improving qualifications, training and involvement of employees in safety activities.

Within this systemic approach to safety management, some provisions are developed, which include:

- information on types of threats and risks associated with the workplace,
- information on how to deal with an accident at work or injury,
- information on the benefits of eliminating dangerous situations and reducing risks,
- the obligation to conform with safety regulations, the description of the employees' responsibilities and competencies as well the possible consequences of any failure to comply with the work safety regulations.

RISKS IN ADHESIVE BONDING OPERATIONS

Mechanical Hazards

Mechanical hazards, e.g., in grinding machines, sanding machines, manual tools for surface preparation, result from the presence of tools (e.g., grinding wheels), instruments and holders, workpieces, ejected solid or liquid materials.

Accidents are sudden and unexpected and may result in staff injuries. In order to monitor and manage the risk caused by mechanical hazards, a work safety analysis as well as fault and event trees are used.

Due to the nature of the activities performed, the following characteristics may influence the occurrence of health hazards during the preparation of objects for adhesive bonding (Letho, 2008):

- shapes - elements with sharp edges, elements difficult to grasp,

- position of the elements in relation to each other - especially during operation - creation of crushing, entanglement and catching zones (e.g., by the grinding wheel),
- mass and stability - due to gravity,
- speed and acceleration,
- insufficient mechanical strength - causing fractures and cracks,
- energy - e.g., pressure of gases or liquids.

Due to mechanical hazards, dangerous situations may occur, which are defined as “cause-effect” (Letho, 2008):

- seizure - pulling in clothing or hands, impact, crushing,
- fracture, chipping, breakage - ejection of a fragment of a grinding wheel, damaged object, contact, impact, abrasion,
- spill - slip, trip, fall, twist, fracture, fall,
- grip - contact, abrasion, impact, injury,
- impact - shock, open wound, amputation, fracture, dislocation, displacement.

Their occurrence may result in material losses or injuries to people.

Electrical Hazards

In adhesive bonding operations, electrical equipment is used. They can be not only grinders, sandblasting machines, hand tools, but also bathtubs for washing, furnaces, guns for heating glue, autoclaves, etc. (Cecińska B., 2004).

Electrical hazards can result in injury or death from electric shock and burns. Electrical hazards can be caused by:

- contact with live parts (normal state),
- contact with parts that are live as a result of a failure (e.g., faulty insulation),
- damaged, inadequate or missing insulation,
- clcctrostatic phenomena,
- short circuit caused by network overload, damage to electric cables, melting of shields, casings, etc., - short circuit caused by overload of the network, damage to electric cables, melting of shields, casings, etc.

An accident caused by electric shock can have further consequences - people falling down, objects being dropped, shock, damage to property.

Thermal Hazards

In preparatory operations, consisting in the creation of a specific condition of the object surface, as well as during the application of an adhesive layer or hot curing, it may be necessary to heat up the etching solution, glue or connection. Thermal hazards may cause:

- combustion, burn, exposure to high temperature objects or materials, explosion, presence of flame, radiation of the heat source,
- health deterioration- as a result of prolonged exposure to a hot working environment.

Noise Hazard

Staying in workplaces where unwanted sound is emitted that hinders or precludes people from communicating, doing their work or resting can cause a variety of problems. The multiplication of noise sources and the increase in its intensity escalate its harmful influence. Noise can cause:

- hearing fatigue, tinnitus,
- progressive deafness and permanent hearing loss,
- fatigue, irritability, stress, attention deficit,
- elevated blood pressure, cardiac arrhythmias,
- balance and consciousness disorders.

Limited possibilities of speech and acoustic communication can cause subjective feelings of insecurity, loss of productivity and quality of work, additional risks.

For instance, using sandblasting equipment in adhesive bonding is an onerous task and a source of noise (Kowalewski S. & Others, 2002).

Chemical and Biological Hazards Arising From Materials and Substances

Adhesive bonding operations require the use of various chemical preparations, e.g., for degreasing, etching or bonding purposes.

Chemical substances in adhesive bonding can be dangerous and harmful to health and the environment. The chemical hazard results from (Niecińska B. & Pacana A., 2012):

(a) the kinds of the effects on the human body and it may include the following substances:
- corrosive – substances which may cause the destruction of living tissues through skin contact,
- irritant - non-corrosive substances which may cause inflammation through short, prolonged or repeated contact with the skin or mucous membrane,
- toxic - substances which may cause serious, acute or chronic health hazards, or even death if they are inhaled, ingested or if they penetrate the skin,
- harmful - substances which may cause a limited health risk if they are inhaled, ingested or if they penetrate the skin,
- carcinogenic - substances which may induce cancer or increase its incidence if they are inhaled, ingested or if they penetrate the skin,
- toxic for reproduction - substances which may cause non-hereditary congenital deformities or increase their incidence if they are inhaled, ingested or if they penetrate the skin,
- mutagenic - substances which may induce hereditary genetic defects or increase their incidence if they are inhaled, ingested or if they penetrate the skin.

(b) the state and possible reactions in the environment:
- explosive - substances which may explode under the influence of high temperature, open flame, shock or are sensitive to friction,
- oxidizing - substances which exhibit strongly exothermic reactions in contact with other substances, in particular flammable substances,
- flammable - substances with a relatively low flash point (below 21°C referred to as highly flammable), liquids, substances which can heat up and eventually catch fire in contact with air at ambient

temperature, gases stored under pressure, as well as substances which, in contact with water or damp air, form highly flammable gases in dangerous quantities,
    - infectious - a combination of chemical and biological threat due to the possible presence of microorganisms in chemical substances - bacteria, viruses, fungi. These organisms can cause human diseases, damage to materials (e.g., plastics, wood) and loss or alteration of their properties, and can produce additional harmful substances (e.g., mycotoxins, allergens),
    - substances which, in contact with water, air or acid, release toxic or highly toxic gases,
    - eco toxic - substances which present or may present an immediate or delayed threat to one or more parts of the environment,
    - substances which after the end of the neutralization process (so, the post-production stage is also important in the context of adhesive waste management) can freely release another substance, e.g., in the form of leachate; this substance may have undesirable properties already mentioned earlier.

Chemical and biological hazards can be created by materials and substances used in the bonding process in machines, resin/hardener mixtures and other additives. Health effects result from inhalation of aerosols, vapors, dusts, contact with the skin and mucous membranes, entering the body through the digestive tract.

Additional risks are posed by supplementary operations in manufacturing processes - transport and storage, which can lead to leakage, fire or explosion, as well as waste management, which also include residues of chemical substances with different properties, sometimes mixed with each other and creating solutions with different characteristics.

Other Hazard: Radiation

Slightly less frequently, as a result of the applied pre-bonding surface treatment technology, there may be hazards caused by radiation.

There are hazards from various sources:

- low frequency,
- radio and microwave frequencies,
- infrared,

- visible light,
- ultraviolet,
- X-rays and γ,
- α and β radiation, electron or ion beam radiation,
- neutrons.

Risk Reduction

In industrial practice, it is impossible to reduce risk to zero, i.e., to achieve absolute safety. Risk should remain at a minimum level, justified by knowledge, technical and economic capacity, awareness and needs of people.

Safety management uses the so-called Triad of Safety, which is based on an iterative (stepwise), three-step process of hazard reduction. The sequential steps in the Triad also illustrate their hierarchy and effectiveness in achieving a safe state (Kowalewski S. & Others, (2002; Kowalewski S. 2002).

The Triad of Safety

I - The most important thing is to eliminate hazards or exposure, which is the same as reducing risk at the design stage of equipment or workstations by making all structures internally safe.

II - If hazards cannot be eliminated due to equipment or workstations (e.g., due to the technology used, when equipment of a specific design and function generates a specific state of hazard) - safety measures must be taken in the form of additional equipment, guards, stops, etc.

III - If, despite the application of safety measures, the residual risk remains so high that, although tolerated, it must be reported accurately to the staff. Sometimes at this stage people are additionally equipped with personal protective equipment: glasses, masks, gloves, protective clothing, etc.

With many technologies, including traditional bonding technologies, it is not always possible to eliminate hazards. Hazards result from the technological requirements of the process, e.g., the solutions used in galvanic baths are the way they are (they are composed of specific chemical substances, toxic, harmful or troublesome chemical compounds) because they exert the expected impact on the surface layer of the products prepared for bonding .

It is then difficult to manage security and implement the first postulate of the Triad.

In such a situation, when it is not possible to apply safe processing methods, other safety **procedures** (principles) are implemented: principles of workstation layout, analysis of safe transport and storage possibilities, principles of ergonomics, proper organization of working time in order to reduce exposure to harmful factors, etc.

Supplementary safety measures, most often in the event of an emergency:

- **devices to prevent current or imminent danger situations**; for gluing the following examples are applicable (Pośniak M., 1999; Zawieska M., 2004; Oczoś K., Kawalec A., 2012):
    - **in case of leakage of** galvanic bathtubs - liquid sensors, sensors of substance concentration, floor shape enabling to stop, collect and remove spilled liquid; such a risk exists in the case of chemical methods of surface preparation before adhesive bonding,
    - **in case of material ignition** during abrasive machining - organizational measures implemented in the vicinity of the machine tool structure (1st step in the safety triad): containers with cast iron chips; this risk exists for the preparation of magnesium alloys or beryllium alloys for bonding by grinding,
    - **in case of damage to containers during transport or storage** - concentration sensors, organizational procedures involving arrangements for visual inspection of the containers and the establishment of procedures for preventive and protective action,
- **emergency stop devices** – in the case of grinders, sanders, furnaces, climate chambers, fans, etc.,
- **devices for rescuing people** - escape routes, shelters, additional manual controls replacing automatic devices, devices changing the state (reversing the direction of movement, changing the temperature or pressure to a safe value, etc.).

Safety procedures should also be established for maintenance services. Devices requiring maintenance and adjustment, especially those frequently used in preparatory operations during bonding, such as: grinding machines, sandblasting machines, furnaces, baths for galvanic baths, should be provided with (Pihowicz W., 2008):

- appropriate access to the controls, mechanisms, electrical connections, liquid supply, etc. - manifested as lack of redundant equipment in the

vicinity and ensuring sufficient space for maintenance workers to perform the overhaul,
- ease of handling, particularly guards and covers, including the possibility of unscrewing and moving them using the required human muscle power,
- a limited number of special tools,
- ease of observation in the context of rapid reaction to danger.

Labeling

Information about residual risks in the workplace should be provided to employees. This can be done in different ways. The description of the workstation should include device data, workstation purpose, and instructions for safe use. In addition, the information should contain separately or collectively the issues of transport, start-up (assembly, installation, adjustment), use (adjustment, programming, changeover, operation, cleaning, maintenance, repair), decommissioning, disassembly, scrapping (Orlenoil, 2020).

In many cases, a symbol is an effective way of communication. Visual management is used for security purposes, and then the graphic symbols – pictograms – are the carriers of a specific message. Pictograms may inform, e.g., about the obligation to wear protective clothing, to put on a shield, to use the specific operating parameters of the device, etc.

In the case of chemicals, safety data sheets shall be used. They should be supplied free of charge by the manufacturer.

ANALYSIS OF THE IMPACT OF TRADITIONAL ADHESIVE BONDING TECHNOLOGY ON HUMANS

During the preparation of adhesive mixes, the preparation of surfaces to be bonded, and during the bonding operation itself, people may be exposed to a variety of odor triggering and toxic and sometimes flammable substances. These are mainly organic solvents, concentrated acids; alkalis and others, e.g., phenol, alkaline metal vapors, alkaline metal solutions in ammonia, molten sodium acetate. Harmful or toxic effects of these substances may occur through the mouth, skin and mucous membranes. In the bonding process the contact with the substance is more likely through the skin and eyes than through the mouth. Eyes and nose are particularly sensitive. The “hypersensitivity” that occurs in some people is most often the result of repeated skin contact with a harmful substance. Then, the way to recover is to avoid contact with such substances. However, chronic poisoning can occur in the case of long-term exposure to harmful substances. They are usually associated with the inhalation of toxic substances and can lead to various diseases and even cancer. The use of these substances requires special care and appropriate working conditions at workplaces (Ptaszyński B., 1994).

A properly made adhesive bond requires a properly prepared surface and a bonding substance, i.e., an adhesive. The process of surface preparation consists of a number of activities, such as (Czaplicki J. & Others, 1987; Rudawska A., 2014):

- degreasing, i.e., removing layers of grease and silicones protecting surfaces against corrosion until the adhesive bond is made; removing layers of dust, particles left after previous machining operations, other organic and inorganic compounds (e.g., fats, previously used adhesives),
- cleaning of the surface in an acidic or alkaline bath followed by a subsequent rinse,
- special treatment - chemical or mechanical - aimed at producing a surface layer with specified properties (e.g., roughness to develop before application of an adhesive), removing corrosion layers, varnishes, paints or other coatings from previous operations.

During the preparatory activities, both the worker and the environment are exposed to harmful chemicals.

Furthermore, this risk does not disappear after surface preparation because the adhesives applied, including the auxiliary substances needed for bonding (hardeners, fillers), due to their chemical structure and certain properties, can also have a negative effect on both the human body and the place where the person is present.

Degreasing and Surface Cleaning

Various substances are used to remove greasy or dirty layers and they have specific ecological characteristics. Examples include TRI (trichloroethylene) which can cause cancer, irritate the eyes and skin and its vapors can cause drowsiness and dizziness (there is a risk that changes in the body are irreversible); and acetone, which is flammable, poses a risk of explosion or fire, irritates the eyes and mucous membranes, dries the skin. Acetone vapors may cause drowsiness and dizziness (LabChem, 2020).

Cleaning the surface with solvents is intended to remove traces of oils and fats which cause poor wettability and adhesion. Cleaning operations are performed by immersing the object in a suitable solvent, preferably heated, or in the vapor phase. Aromatic hydrocarbons, aliphatic hydrocarbons, chlorinated aliphatic hydrocarbons and ketones are used.

In turn, immersion in acid baths is used to dissolve oxides, hydroxides and other compounds in inorganic acids. Most commonly used are concentrated hydrochloric acid, concentrated sulphuric acid, concentrated nitric acid, and hydrofluoric acid. It is also possible to use mixtures of these acids and chromium mixture (concentrated H_2SO_4 + $K_2Cr_2O_7$ at 70°C). However, immersion in alkaline baths requires the use of strong alkalis and soda.

Special baths are used to prepare the surface of objects made of materials which are difficult to stick together, e.g., alkaline metal solution in ammonia, molten sodium acetate, alkaline metal vapors (Cagle, 1997).

Digestion, Pickling, Anodizing

Digestion, pickling, anodizing are chemical surface preparation operations requiring a special bath of various substances. These may include acids of various concentrations (e.g., sulphuric - dangerous, causes burns to the skin, eyes, digestive system, respiratory system), alkalis (e.g., NaOH - strongly corrosive, causes severe burns), or chromates (e.g., sodium dichromate) - considered carcinogenic and mutagenic substances, very toxic by inhalation, very dangerous in the case of long-term exposure, corrosive, causing burns or sensitization in contact with the skin or while inhaling.

Bonding

In order to make adhesive joints between metals and metals, many types of adhesive compositions are produced and used, which are mixtures of various chemical compounds (resins, adhesives, hardeners, modifiers, sealants, fillers and others) posing a specific threat, including:

- epoxy resins - which are not toxic, but may cause sensitization, manifested by skin changes; styrene absorbed by the respiratory route is harmful (mutagenic, possible carcinogenic effect), irritates the eyes and skin, leads to disorders of the digestive system, kidneys and digestive system,

- cyanoacrylate adhesives - whose vapors are not toxic but can irritate eyes and mucous membranes,
- phenolic resins, phenol-formaldehyde resins - based on toxic phenol and formaldehyde, poisonous substances suspected of mutagenic effects, causing acute poisoning, respiratory disorders, nervous system disorders up to the loss of consciousness, in milder cases of poisoning they cause a sore throat, skin inflammations, orientation disorders (similarly, adhesives based on polyvinyl acetals modified with phenol-formaldehyde resin),
- adhesives based on polyvinyl chloride - they are flammable and toxic due to the content of volatile and flammable solvents, e.g., acetone, toluene and others,
- methacrylate adhesives - causing headaches, allergies and skin irritations,
- polyurethane adhesives - flammable and toxic due to solvent content (acetone, benzene, toluene), some, e.g., polyisocyanates dangerous due to rapid reaction with the skin, irritating to eyes, respiratory tract, there is limited evidence of carcinogenicity. Breathing in vapors and aerosols can lead to necrotic changes and consequently to bacterial infections of altered tissues (Pośniak M., 1999).

However, hardeners containing amine compounds, which can cause skin problems ranging from allergic lesions and rashes to persistent eczema and skin inflammation, are toxic to a large extent. In the case of many glues, e.g., for wood, dilute hydrochloric, sulfuric or orthophosphoric acid is used. Extremely dangerous is e.g., hardener Z1 (tri-ethylene diamine) with corrosive and vaporizing action, in liquid and volatile state (Czaplicki J. & Others, 1987).

Measurements

Measurements of the wettability of surfaces are needed before adhesion; some methods require the use of other substances in addition to water. Often encountered formamide - highly toxic, requires work only under the fume extractor, causes gastrointestinal problems, irritates the skin and eyes, vapors or mists irritate the mucous membranes and upper respiratory tract, in the event of prolonged exposure it can cause severe burns and blisters and cause reproductive disorders; and diiodomethane, which is irritant to skin and eyes, is harmful when absorbed in the respiratory tract (inhalation causes drowsiness and even leads to unconsciousness) and, if swallowed, is believed to be carcinogenic.

Threats to human health can be serious. Poisoning with various substances can be acute or chronic when the symptoms persist for many years. Poisonous substances enter the body especially by inhalation, but also through the digestive tract, skin surface and eyes.

Table 1 presents examples of health consequences resulting from contact with various chemical substances (Czaplicki J. & Others, 1987).

Table 1. Health consequences

Type of substance	Health problems
Irritant poisons	Irritation of respiratory tract mucous membranes, eyes and skin (acidic vapors and gases, ammonia)
Drug poisoning	Effects on the central nervous system, cerebral cortex (e.g., aliphatic hydrocarbons); irritation of the nervous system (carbon disulphide, alcohols); harmful effects on the hematopoietic system, especially bone marrow (aromatic hydrocarbons)
Carcinogenic substances	Destruction of bone marrow, leukaemia (benzene), lung cancer (nitric acid)
Mutagenic substances	Changes in genetic material (benzene, nitric acid)
Substances causing skin and eye diseases I - concentrated inorganic acids	Denaturation of the protein substance, skin necrosis, deep changes associated with skin carbonization (concentrated sulphuric acid)
Substances causing skin and eye diseases II - Concentrated lyes	Saponification of fats and dissolution of creatine, penetration into the skin. Bleeding, hard to heal wounds, deep scars
Phototoxic substances	Lesions occurring after the contaminated site is illuminated (toluene, xylene)

The possibility of eye burns is an underestimated danger. Especially in the process of glue preparation, surface cleaning and gluing, there is a possibility of contact with vapors of organic solvents, acid vapors and gases, ammonia. These substances can cause inflammation and swelling of the eyelids and conjunctivitis and cornea by irritating effects. Such inflammations may be mild in the case of short-term presence in a room with poor ventilation, but also acute or chronic, as a result of prolonged presence at the workplace or in a significantly polluted air. Eye burns may result in permanent loss of vision or other illnesses. The most dangerous are alkalis (alkaline hydroxides, ammonia), because they cause tissue necrosis. Concentrated inorganic acids in turn cause dehydration and denaturation of corneal and conjunctival epithelial protein (e.g., HCl, H_2SO_4, HNO_3). Organic solvents cause damage to the epithelium of the cornea and conjunctiva.

In the case of chemical substances there is a phenomenon of synergy – reinforcement – mixtures of substances can be more harmful than their individual constituents separately. Alcohol, for example, increases the rate at which vapors and gases are absorbed. For this reason, organic solvents, hydrocarbons, chlorinated hydrocarbons, carbon disulphide, etc., are more easily absorbed by the body, have a stronger effect and give symptoms of illness or poisoning in a shorter time (Czaplicki J. & Others, 1987).

ANALYSIS OF THE IMPACT OF TRADITIONAL ADHESIVE BONDING TECHNOLOGY ON NATURAL ENVIRONMENT

Various substances used in adhesive bonding operations can not only be a threat to human health, but can also be a significant environmental aspect.

I - demand for chemical substances

Solvents are an example. They can be used for adhesives - when aliphatic hydrocarbons, aromatic hydrocarbons and chlorinated or fluorinated aliphatic hydrocarbons are used. Aliphatic hydrocarbons are volatile products of crude oil distillation and are easily flammable (gasoline). Aromatic hydrocarbons are better solvents but also more toxic (most commonly used in this group are benzene, toluene and xylene). Acyclic hydrocarbons, on the other hand, are chlorinated or fluorinated, are good solvents and are more difficult to ignite, but they are toxic compounds. They are also not resistant to light and heat. In unfavorable conditions they may separate hydrogen chloride or chlorine (this group includes dichloroethane, trichloroethane, trichloroethylene, methylene chloride, trifluoroethane). In addition, cyclohexane, ethyl acetate, carbon tetrachloride, dioxane, amyl acetate, ketones, esters, carbon disulphide can be used (Ptaszyński B., 1994).

They can be listed according to their toxicity, starting from the weakest: gasoline, cyclohexane, xylene, toluene, benzene, dichloroethane, carbon disulphide.

II - demand for electricity

Some of the adhesive bonding operations require a temperature-controlled environment and other special devises powered by electricity.

III - emissions to air

Many different chemicals are used in the process of bonding. For this reason, they are an important source of air pollution in workstations and production halls when released into the environment. Harmfulness of these substances occurs when they reach a certain concentration in air. The permissible concentrations are defined by standards (NDS, NDCh).

In recent years, great attention has been paid to ensuring the environmental safety of the technologies used and environmental monitoring has been applied. Also, air pollution by chemical substances may be the subject of air pollution monitoring. Air samples are taken from the employee's environment and analyzed in the context of compliance with the requirements of the standards, detection of the so-called emitters and the impact of odors and toxic substances on human beings and the environment (Czaplicki J. & Others, 1987).

The importance of such a study can be demonstrated by the example of benzene (used as a solvent) as one of the most dangerous industrial poisons. Benzene (C_6H_6 - aromatic hydrocarbon) is mainly absorbed by the respiratory route. Its vapors absorbed through the skin do not give rise to drastic problems. However, the inhalation of vapors, even at low concentrations and in a short period of time, can cause severe illnesses. They can take the form of changes in the nervous system; affect the blood and hematopoietic system. Vapors in high concentrations may irritate the mucous membranes of the respiratory route, nose and eyes. Long-term inhalation may cause drowsiness, exhaustion, headaches and dizziness, nausea, heart rate acceleration, even up to loss of consciousness. In extreme cases, seizures, nerve paralysis and death can occur. Benzene is a sneaky poison. Chronic poisoning may develop slowly and may not give immediately visible symptoms. Sometimes workers complain about growing weakness, followed by changes in the blood composition, decreased clotting, and bleeding from the nose and gums. The marrow is damaged, resulting in anemia and leukemia.

The example of benzene illustrates the dependence of human health on the quality of the environment, in particular production conditions - the working environment.

IV - waste load

In the context of environmental risks, waste generation must also be taken into account, analyzing successively the operations related to the preparation of surface and tools as well as to the cleaning of bonded surfaces.

Many different types of waste are produced at the bonding stations.

V - danger of ignition and explosion

In addition to health risks, the chemical substances used in gluing operations pose a risk of ignition and explosion.

Flammable and explosive liquids are used in production halls. There, it is somewhat easier to supervise the environmental conditions due to the presence of people. They can also be stored in warehouses where staff presence is less frequent. Then, it is necessary to ensure appropriate storage conditions and to monitor those conditions by technical means. Safety data sheets can provide information on substance properties.

ANALYSIS OF THE IMPACT OF LASER PROCESSING

A laser is a device that produces or amplifies a coherent beam of electromagnetic radiation in the spectrum from ultraviolet to far infrared (there are now also lasers that produce coherent X-rays and microwaves). In lasers, the phenomenon of radiation amplification due to controlled emission forced in the active medium (laser substance) after reversal of planted energy levels was used.

Due to the characteristics of the laser beam described in Chapter 6 and its wide range of applications, the potential hazards in working with lasers must be considered for three reasons:

1) Danger due to impact of the beam on the human body
2) Health and environmental aspects resulting from the impact on the materials processed
3) Environmental aspects resulting from the use of equipment

Dangers Resulting From the Exposure of the Human Body to Laser

Hazards from laser radiation concern the eyes and skin and include accidental short-term exposure. In such cases, medical applications and deliberate exposure to radiation shall not be taken into account in the occupational risk assessment of lasers.

The most vulnerable to radiation are the eyes. For wavelengths from 400 nm to 1400 nm (the range of the most common lasers), the greatest danger is retinal damage. Such radiation penetrates the eye and is focused on the retina.

Radiation of wavelengths below 400 nm and above 1400 nm does not penetrate into the eye, but causes damage to the cornea of the eye (Oczoś K., 1988).

For the skin, much higher doses of radiation, in comparison with the eyes, cause biological consequences of the laser radiation impact. In the case of skin, exposure to the beam may cause charring, burning or erythema on the skin.

The effects of exposure to laser radiation are shown in Table 2 (Oczoś K., 1988).

Laser beam area marking is required – Figure 2.

Table 2. Harmfulness of exposure to laser radiation

Radiation	Effects	
	Eyes	**Skin**
Ultraviolet C (180-280 nm)	Inflammatory damage to the cornea	Erythema, accelerated skin aging, increased pigmentation
Ultraviolet B (280-315 nm)		
Ultraviolet A (315-400 nm)	Photochemical cataract	Pigment darkening, photosensitization reactions, burns
Visible radiation (400-780 nm)	Photochemical and thermal damage to the retina	Photosensitization reactions, burns
Infrared A (780-1400 nm)	Cataracts, retinal burn.	Burns
Infrared B (1.4-3.0 μm)	Corneal fibrosis, cataract, corneal burn.	Burns
Infrared C (3.0 μm -1 mm)	Corneal burn	Burns

Figure 2. Determination of the impact zone of a concentrated energy beam

Additional protection: screens, housings, glasses – Figure 3 - are especially required when working on glossy materials with a very smooth surface.

Figure 3. Photographs of safety devices used in laser processing: a) glasses, b) double extractor for air filtration, c) modern equipment housings with impermeable glass

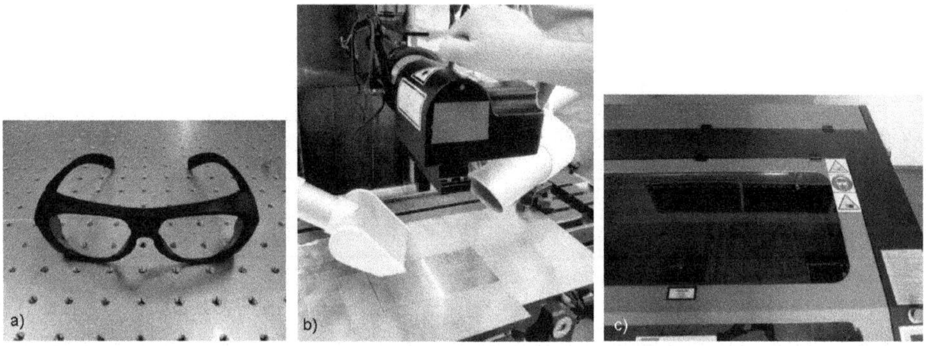

Health and Environmental Aspects Resulting From the Impact on the Materials Processed

Metalworking

In trace amounts, metallic elements are needed for proper functioning of the human body. However, the problem starts when the limit values are exceeded under industrial conditions. Metals, which enter the human body by inhalation, in the form of dusts (aerosols of solid particles) or fumes (aerosols of liquids), accumulate in organs, brain and bones, leading to chronic diseases. Lead, mercury, chromium, iron, zinc, tin are examples of metals that interfere with the metabolism of the body, can have toxic or carcinogenic effects. In category 1 (according to the classification of the European Union, Directive 67/548/EEC), i.e., in the group of metals with a proven carcinogenic effect, there are arsenic and its compounds, beryllium and beryllium compounds, chromium (VI) compounds, cadmium and cadmium compounds, nickel salts. The effects of metal accumulation can only be felt after many years of work, even at retirement age, when symptoms of illness are not associated with poor working conditions in the past. Flu-like symptoms may occur when inhaling aluminum alloy vapors and zinc coatings. Metals in the air can be found in various mixtures and fractions (Pośniak M., 2001).

Welding aluminum alloys, among other things, poses some difficulties in laser processing, because the material has a low absorption capacity and a large part of the radiation is reflected from the surface of the object and dispersed in the environment. For this reason, high power lasers are used to weld low-absorbent materials. In addition, welding stations, generally specially designated, should have screens to protect the health of personnel.

Processing of Plastics

Laser processing of plastic products can be interesting because of the possibility of shaping, coloring or other specific properties, such as in 3D techniques, or material saving and easy process control, such as laser cutting. However, not all plastics are safe for humans as a result of their chemical structure. In many cases, the causes of the danger are heating operations that release emissions of harmful substances. An example is polyvinyl chloride, which after adding plasticizers, dyes, fillers, is a source of emissions of chlorine compounds, benzene, carbon monoxide and aromatic hydrocarbons. These substances cause changes in the respiratory system, irritation, lung diseases, asthma, cancer, may have a narcotic effect, cause excessive sleepiness or excitement. Due to the fact that most manufacturing processes of plastic products require the addition of other substances: catalysts, surfactants, blowing agents, fillers, pigments, plasticizers, flame retardants or oxidizers, it should be borne in mind that these chemical compounds - harmful and dangerous - may also be inhaled by humans during processing operations (Pośniak M., 2001).

The problem of substance emission may be significant. For example, cleaning the surface with a laser beam by removing layers shows that the evaporated matter can contain hazardous chemicals such as paint, heavy metals, aromatic hydrocarbons and others (Walker, 2002).

This problem can be avoided by using the right equipment for laser processing stations. The processing zone should have an extractor of sufficient power and the extracted air should pass through an appropriate filter. Nowadays, manufacturers offer various technical solutions and it is not a problem to choose the right filter even for mobile laser heads (Figure 4).

Figure 4. Personnel protection: extractor hood - mobile filters - in case of emission into the air, screens for bystanders in case of reflected radiation

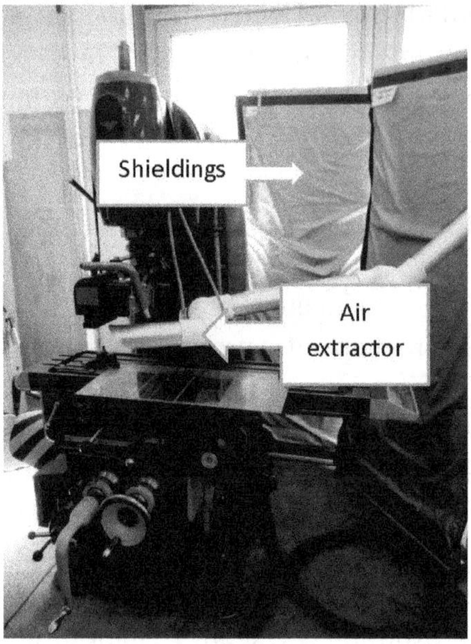

Figure 5. Eye safety: a) radiation invisible to the eye, visible in the picture, b) choice of protective goggles according to wavelength

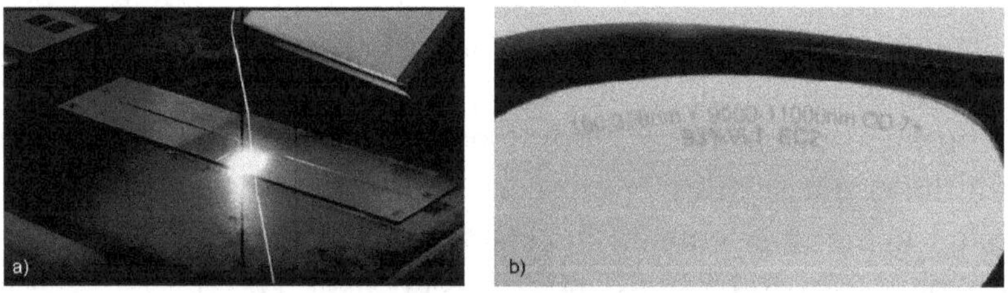

CONCLUSION

From the analysis of the health and environmental aspects of bonding technology, it appears that traditional methods, especially surface preparation stage before applying the adhesive compound pose a number of problems. The issues concern the negative impact of chemical substances on human health, as well as environmental impact due to hazardous waste as a by-product of production, waste and liquids often contaminated with toxins, toxic fumes emissions, the risk of fire or explosion.

The use of lasers makes it possible to significantly reduce these nuisances and problems, especially due to the lack of many chemicals. However, it is important to be aware that any change in technology gives rise to risk reassessment. Such measures are justified in this specific case as it is not possible to eliminate all the harmful effects. Laser processing, which is beneficial for many of the reasons described earlier, is also associated with certain inconveniences.

REFERENCES

Badida, M., Ladomersky, J., Kralikova, R., Moravec, M., & Liptai, P. (2014). *Environmentalne inzinierstvo* [Environmental engineering]. Kosice: ELFA, s.r.o.

Cagle, Ch. V. (1977). Kleje i klejenie [Glues and gluing]. Warszawa: Wydawnictwa Naukowo-Techniczne.

Chaaban, M. (2001). Hazardous waste source reduction in materials and processing technologies. *Journal of Materials Processing Technology*, *119*(1-3), 336–343. doi:10.1016/S0924-0136(01)00920-7

Ciecińska, B. (2004). Przygotowanie i organizacja produkcji [Preparation and organization of production]. Laboratorium. Rzeszów: Oficyna Wydawnicza Politechniki Rzeszowskiej.

Ciecińska, B., & Pacana, A. (2012). Podstawy zarządzania środowiskowego [Fundamentals of environmental management]. Rzeszów: Oficyna Wydawnicza.

Czaplicki, J., Ćwikliński, J., Godzimirski, J., & Konar, P. (1987). Klejenie tworzyw konstrukcyjnych [Bonding of construction materials]. Warszawa: Wydawnictwa Komunikacji i Łączności.

Durlik, I. (2007). Inżynieria zarządzania. Strategia i projektowanie systemów produkcyjnych [Management engineering. Strategy and design of production systems]. Warszawa: PLACET.

Editorial. (2012). Industrial and hazardous waste management. *Journal of Hazardous Materials*, 207-208.

Geng, H. (2004). *Manufacturing Engineering Handbook*. New York: McGraw-Hill.

Groover, M. (2008). *Automation, Production Systems, and Computer-Integrated Manufactirung*. Pearson Education Inc.

Kobyłko, G. (2000). *Proekologiczne zarządzanie przedsiębiorstwem [Pro-ecological company management]*. Wrocław: Wyd. Akademii Ekonomicznej.

Kowalewski, S. (2002). Podstawy bezpieczeństwa maszyn [Fundamentals of machine safety]. Warszawa: CIOP.

Kowalewski, S., Dąbrowski, A., Dąbrowki, M., Pietrzak, L., & Wroński, S. (2002). Charakterystyka zagrożeń stwarzanych przez maszyny produkcyjne [Characteristics of hazards posed by production machines]. Warszawa: CIOP.

LabChem. (2020). *Acetone safety data sheet.* https://www.labchem.com/tools/msds/msds/LC10420.pdf

Letho, M., & Buck, J. (2008). *Introduction to human factors and ergonomics for engineers*. New York: Taylor & Francis Group.

Lewandowski, J. (1998). Zarządzanie jakością. Jakość, ergonomia, bezpieczeństwo pracy, ochrona środowiska [Quality management. Quality, ergonomics, work safety, environmental protection]. Łódź: Marcus.

Oczoś, K. (1988). Kształtowanie materiałów skoncentrowanymi strumieniami energii [Shaping materials with concentrated energy streams]. Rzeszów: Wydawnictwo Politechniki Rzeszowskiej.

Oczoś, K., & Kawalec, A. (2012). Kształtowanie metali lekkich [Light metal shaping]. Warszawa: PWN.

Orlenoil. (2020). *Chemical safety data sheet*. Retrieved from https://www.orlenoil.pl/PL/NaszaOferta/KartyCharakterystyki/KartyCharakterystyki/1246.pdf

Pihowicz, W. (2008). Inżynieria bezpieczeństwa technologicznego [Technological security engineering]. In Problematyka podstawowa. Warszawa: WNT.

Podgórski, D. (2001). Wytyczne integracji systemów zarządzania bezpieczeństwem i higieną pracy z systemami TQM [Guidelines for integrating occupational health and safety management systems with TQM systems]. CIOP.

Pośniak, M. (1999). *Zagrożenia chemiczne w wybranych procesach technologicznych – cz.1 [Chemical hazards in selected technological processes]*. Warszawa: CIOP.

Pośniak, M. (2001). *Zagrożenia chemiczne w wybranych procesach technologicznych – cz.2 [Chemical hazards in selected technological processes].* Warszawa: CIOP.

Ptaszyński, B. (1994). *Niektóre problemy BHP w technologii klejenia [Some OHS problems in gluing technology]. In Kleje i technologia klejenia.* Łódź: Wyd. Pollitechniki Łódzkiej.

Rudawska, A. (2014). Selected aspects of the effect of mechanical treatment on surface roughness and adhesive joint strength of steel sheets. *International Journal of Adhesion and Adhesives*, *50*, 235–243. doi:10.1016/j.ijadhadh.2014.01.032

Walker, C. H., Hopkin, S. P., Sibly, R. M., & Peakall, D. B. (2002). Podstawy ekotoksykologii [Basics of ecotoxicology]. Warszawa: PWN.

Zawieska, M. (2004). Ocena ryzyka zawodowego. 1. Podstawy metodyczne [Risk assessment. 1. Methodological basics]. Warszawa: CIOP PIB.

Chapter 8
Prognosis for the Future for Laser Technology:
Cleaner Production, BAT, and Industry 4.0

ABSTRACT

This chapter presents a modern approach to production management based on actions aimed at minimizing its impact on the environmental. The philosophy of cleaner production was characterized, the idea of which is to prevent pollution, for example, by introducing changes in production technology or reusing production waste. The concept of BAT, the best environmentally effective production methods, was described. In this context, the concept of implementing laser processing as a method to prevent environmental pollution occurring in conventional product bonding was presented. In addition, a discussion on the prospects of implementing a new laser technology in bonding was presented. The developmental conditions visible in the Industry 4.0 revolution, automation, robotization, just in time, and lean manufacturing philosophy, which are conducive to the spread of lasers in place of already known but problematic manufacturing technologies, were demonstrated.

DOI: 10.4018/978-1-7998-4634-5.ch008

THE PHILOSOPHY OF CLEANER PRODUCTION AND BAT

Enterprises may react differently to the external stimuli affecting them in terms of environmental protection. These reactions depend on various factors: the kind of industry, the intensity of the influence of a given factor, thc sizc of the plant, the location, the strategy for dealing with the environment. At present, there are four strategies for dealing with the natural environment (Bernaciak A., Gaczek W.M., 2001):

1. **Passive Strategy**: inactive, neutral - the company observes environmental protection regulations, does not pay penalties and maintains its positive image in the local environment, but does not take any additional actions.
2. **Offensive Strategy**: active - the company uses all means to improve production processes in terms of environmental protection, implements innovative technological solutions, reduces the negative impact on the environment, raises the ecological awareness of the staff.
3. **Innovative Strategy**: active - the company seeks new technologies, constructions and products adapted to the requirements of environmental protection, builds closed water circuits, recycles waste, works on radical improvement of the company's relations with the environment.
4. **Defensive Strategy**: the company withdraws harmful products from the market, renounces technologies that do not meet ecological criteria, improves and modifies processes through partial recycling, partial elimination of environmental damage (passive attitude).

Currently, on a global scale, the postulates of pollution prevention are implemented, i.e., the use of practices, manufacturing techniques, materials, products, services and energy in order to avoid, limit and supervise the generation of emissions, discharges of all kinds of pollutants and waste, in order to reduce the negative impact on the environment. The most desirable is reduction "at the source'', i.e., changes in processes, products, efficient use of resources, substitution of materials and energy, reuse, recovery, recycling, regeneration and processing (ISO 14001, 2015). The use of different ways, according to thc philosophy and capabilities of the plant, is shown in Figure 1:

According to this philosophy, the activities undertaken in the company should have their own hierarchy, according to which the preventive activities should be given a priority (Groover M., 2008):

Figure 1. Pollution prevention philosophy

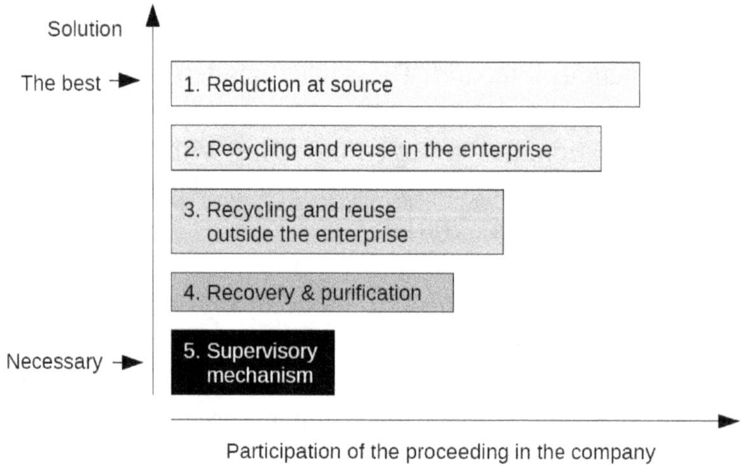

1. **Prevention at Source**: it is possible at the stage of product design, manufacturing process, technological process. It is implemented by substituting harmful materials with safer ones, e.g., abandoning the use of solvent-based paints and replacing them with water-thinnable paints, substituting acid or alkaline baths with other surface treatments not involving chemicals, etc.
2. **Recycling and Reuse of Materials in the Organization**: it is possible at the stage of production, when semi-finished or defective products found during the production process can be reprocessed, e.g., cullet, metal waste in the process of remelting, use of closed water circuit, solvent recovery.
3. **Recycling and Reuse of Materials Outside the Organization**: is realized by transferring waste to other plants in order to process it, e.g., sawdust for the production of chipboards, plastic waste for the production of the items with no specific color requirements.
4. **Recovery and Purification**: visible in the plant practices: waste segregation, energy and water recovery, purification of processing fluids, gases, air filtration from dust, etc.
5. **Supervision**: organizational activities that help in achieving environmental objectives: monitoring the condition of equipment, concentration measurements, modification of pollution composition, organizing a safe place to store, transport, cleaning filters, maintaining machinery and equipment in working order, etc.

The last possibility - called "end of pipe" action - should be used as the last possible option, i.e., after considering actions 1-4.

In the environmental impact assessment, the following factors are analyzed (ISO 14001, 2015):

- emissions to air - dust, gases, noise, heat, odors,
- solid and liquid wastes,
- water consumption and waste water,
- products,
- raw materials used,
- energy sources used,
- wrappings,
- relevance to the local environment: location of the enterprise, location of machinery, warehouses, landfills,
- relevance to safety: the safety of workers and the environment, in the context of an emergency situation which may result in the release of substances into the environment, contamination of the site, groundwater, danger to human health or life.

Entrepreneurs are encouraged to use effective and advanced ways and methods of doing business, the so-called **BAT** (Badida M. & Others, 2014):

- **Best**: most effective in achieving a high level of environmental protection
- **Availabl**e: developed on such a scale that it possible to implement them in economically and technically justified conditions
- **Techniques**: technologies, ways of designing installations, building, maintaining, operating and decommissioning facilities

There are many methods for assessing the environmental impact of technologies. A simple way to do this is an INPUT-OUTPUT analysis, which lists all the factors present in the production process that may have a potential impact on the environment. An example of such an analysis for adhesive bonding operations is given in the Table 1.

In the United Nations Environment Programme (UNEP), the following name was adopted:

Table 1. INPUT-OUTPUT analysis of the adhesive bonding process

Process: Adhesive bonding operation		
Inputs		**Outputs**
· adhesive material · adhesives · hardeners · primers · acetone · gasoline · cloths · harness · stove (autoclave) · agitator tips for glue mixers · brushes · ultrasonic cleaner · sandbox · grinding units/tools · measuring liquids · water · electricity · solutions of acids, alkalis for surface treatment	→	Waste: · used corrundum · adhesive packaging (metal, plastic, glass contaminated with chemicals) · hardener packaging (as above) · used tips (plastic) · dirty fabrics · adhesive residues · hardener residues · contaminated liquids · dirty brushes, spatulas, etc. Surrounding: · the risk of a spill, · dust and vapors emissions to air · odors · noxiousness for people · safety risk in case of an accident (ignition, fire, explosion)

Cleaner Production: an environmental protection strategy based on continuous, preventive, integrated action with regard to processes, products and services, aimed at increasing the efficiency of production and services and reducing the risk for people and the environment.

Cleaner Production is represented by:

- restraint - shaping new attitudes among people, different than the existing ones, and shaping a different hierarchy of values (ecological philosophy),
- TQM - Total Quality Management - quality, not quantity,
- integration - harmonizing human needs with the nature's capabilities,
- saving resources on a local and global scale,
- environmental policy - declaration in which entrepreneurs present actions taken to reduce negative impact on the environment,
- public participation,
- sustainable development - development taking into account the needs of present and future generations,
- equal opportunities for people all over the world.
- The company's Cleaner Production activities are related to:
- production processes - saving water, raw materials and energy, elimination of toxic substances, reduction in waste amount and its negative effect,

- products - reducing negative impacts throughout the life cycle,
- services - introducing environmental issues into design and consumption.

Cleaner Technologies are manifested in:

- using as little raw materials as possible, but these should be renewable, non-toxic and biodegradable raw materials,
- minimum water requirements,
- minimum emissions to air, soil, dust, sewage and waste,
- use of closed substance circuits in processes,
- internal recycling - reuse of waste in the plant.

As regards **Cleaner Products**, they are:

- harmless to ecosystems,
- energy-saving, material-saving,
- produced from renewable raw materials,
- even if produced from non-renewable raw materials, they are recyclable,
- can be disassembled, repaired,
- durable, reusable,
- minimally packaged for distribution,
- appropriately labeled.

The CP strategy in the life cycle of the production process is illustrated by the following diagram (Figure 2):

CP is based on the following principles:

I - waste prevention - make sure that no waste is produced in the production process, use cleaner technologies and cleaner products.

II - if there is no possibility to eliminate waste, make it suitable for recycling and processing

III - as a last resort - if you do not manage to fully recycle your waste, you may landfill it, provided that you ensure environmental safety with regard to noxiousness and health hazards

Waste minimization strategies may vary. Changing environmentally damaging technology is the best preventive approach. Then, you can also meet the requirements of CP (Figure 3).

Figure 2. CP philosophy in the production cycle

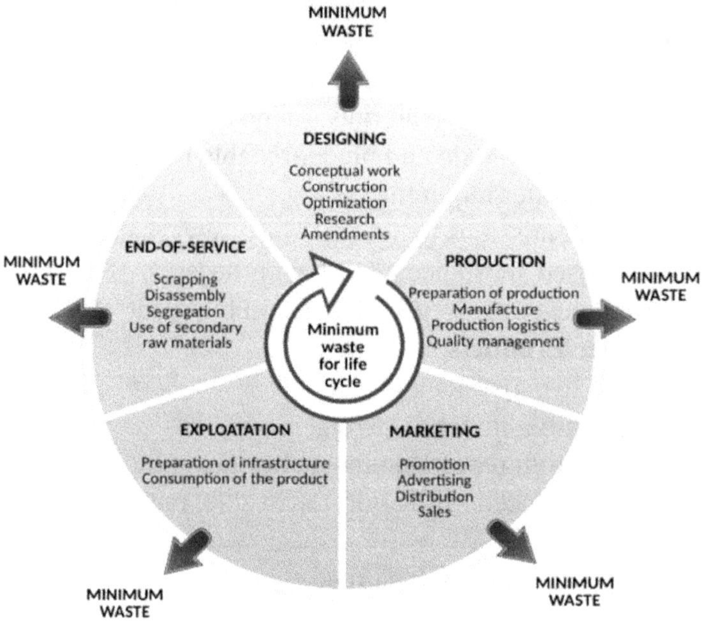

A good example of such a noxious technology is the bonding technology described in the monograph, where the traditional surface preparation operations can be substituted with other environmentally friendly methods such as laser processing.

The implementation of the Cleaner Production Programme brings tangible benefits to both the company and society:

- financial savings - fewer raw materials purchased, more efficient and accurate technologies, no or less waste,
- improved efficiency - a reduced number of purification processes or complete elimination of purification processes,
- the reduction of environmental pollution caused by waste,
- to improve the quality and safety of products,
- reduced risk of accidents involving hazardous chemicals,
- increased safety of workers,
- preservation of the elements of the environment in good condition.

Figure 3: Strategies for waste minimization; position of laser surface preparation technology instead of traditional adhesive bonding processing

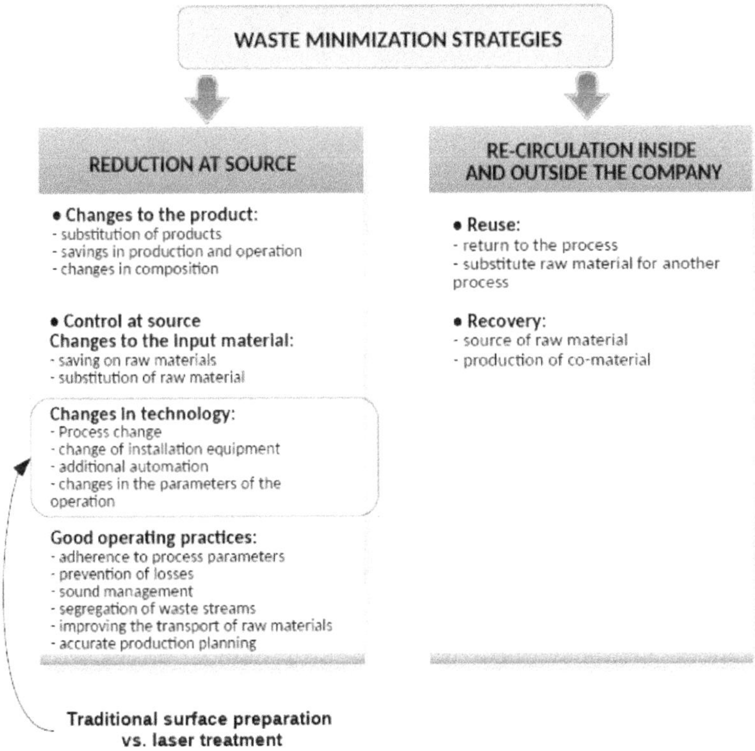

DEVELOPMENT TRENDS OF PRODUCTION PROCESSES

Automation and Robotization

The emergence of flexible production systems was a response of the industry to the increasing fluctuations in demand, stronger competition and growing demands of customers.

The flexibility of the production system is one of the most important features of a market economy and can be achieved through various means, including:

- shaping the assortment of products,
- various construction and technological solutions,
- selection of production equipment,

- selection of the organizational and spatial structure of the production system,
- selection of the method of controlling the flow of materials and products,
- appropriately shaped relations between employees (internal relations) and relations of the enterprise with the environment (external relations) (Lis S. & Others, 1994).

Flexible production systems are now the basis for the functioning and development of modern enterprises. In flexible systems, the implementation of technological processes relies on **automation**, based on program control, which includes, among others, robots, machine tools and numerically controlled technological equipment, automated warehouse system. The second element is IT systems designed to support engineering, design, prototyping, calculation, simulation, process optimization, as well as to support office work (generating reports, technological documentation, tool lists, material balances, measurement cards, etc.) and management (decision support programs, Human Resources management).

The realization of flexible production, automatic and robotized, are machine tools. The basic direction of development of means of production is numerical control. It requires improvement of drives, sensors, measuring devices, so that the operation of the devices runs smoothly, according to an exchangeable program. Considerable changes have taken place in the construction of machine tools. The scope of machining operations has been extended and the range of auxiliary activities has been broadened: manipulation (mounting and unfastening of parts, tools, heads), diagnostics, control and measurement, waste disposal. It has become possible to perform parallel machining and auxiliary activities, which has resulted in an increase in productivity (Honczarenko J., 2000).

The second direction of development is robotization, currently visible on automatic assembly, welding, welding and painting stations. The use of robots ensured a radical reduction in the number of defective products, the elimination of noxious or dangerous work performed by people, the implementation of various types of production tasks in any order.

Progress in the construction of machine tools entails progress in the construction of tools. Manufacturers care about durability, interchangeability, the use of modern materials for tooling and equipment.

Automation should also cover the transport and storage system of products. Appropriately labelled components are handled, stacked and stored in high bay racks thanks to computer-controlled manipulators and stackers integrated into the entire production management system. A modern solution are unmanned trucks, moving along a given trajectory and communicating by means of modern locating devices with traffic controllers.

Modern, automated production systems also include other devices: measuring machines, washing machines, dryers, maintenance stations, packing stations and many others. Many solutions, such as laser devices, have a high potential for automation.

In addition to the development of equipment, it should be mentioned that IT development is essential for the improvement of modern production. It concerns software, networking and other ways of data transmission, taking into consideration the well-known concept of virtual factory, where employees can be located in different places of the globe and the outcomes of their work are transmitted to the parent company's server. In the world of computer science, the progress in miniaturization of computer hardware, its reliability, mobility, access to cheap peripherals is particularly visible. Graphic terminals, mass storage, data collection and storage devices, voice and video input are used. The important aspects are resistance to interference, low price, but also adequate computing power, memory size. In addition to upgrading computer hardware, there is a rapid development of software. Newer, more effective operating systems are being developed, enabling intuitive, multi-programme work based on distributed knowledge bases and object-oriented programming (Lis S. & Others, 1994).

Automation and Robotization of Laser Processing

The construction of laser equipment is conducive to automation and robotization of workstations and enables three-dimensional processing of products. Various laser head configurations are used in the industry, depending on the manufactured assortment (part size, shape complexity, and number of necessary machining operations).

There are at least four types of organizational solutions for robotic laser processing stations (Oczoś K., 1988):

I - the robot controls the movement of the object, the laser beam is stationary, typically focused. An object is held in a robot manipulator and moves within a specific work area, taking a repeatable position. There can be a robot of any structure, computer-controlled in 5 axes: 3 shifts and 2 revolutions.

II - the robot controls the movement of a flexible laser beam delivery system (e.g., fiber optic laser) - the robot manipulator holds the laser head with the optical system, properly aligned. The workpiece is placed on a work table and remains stationary during machining. In this configuration, the most frequently used are articulated arm robots, gate robots and computer-controlled robots.

III - robot with built-in laser beam delivery system - laser head is an integral part of the robot. It is more advanced than type II, with 3-5-axis gate robots and specialist articulated arm robots.

IV - hybrid systems - a combination of a Type I or Type II solution with a different type of treatment. This solution is shown in the figure, where the robot manipulator takes consecutively the objects placed in the workstation zone, places them according to the program in different devices (here alternately in the laser and bending machine).

Figure 4. Robotized picking station

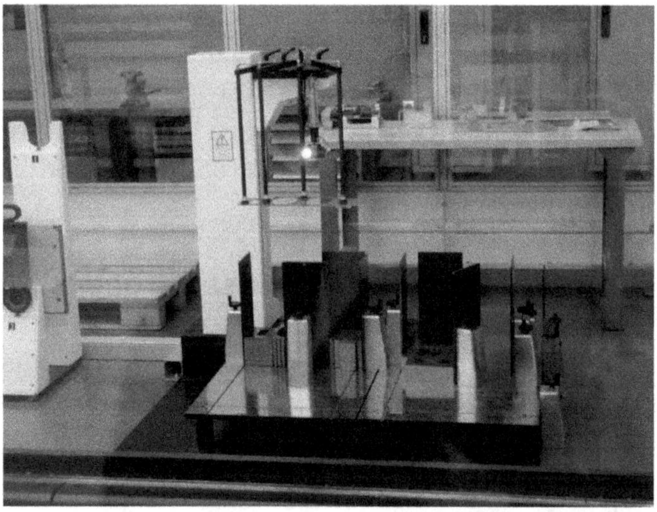

Other technological solutions consist in building a laser head together with an optical system into the structure of cutting machines. You can meet CNC machining centres where workpieces are machined and the laser is used for surfacing, spraying coatings, metallization, etc. Frequently, a second laser for measurements (with or without a touch probe) is built into such machines.

Table 2. Possibilities and limitations of robotized laser stations

Type of station	Capacities	Restrictions
I	- low cost - possibility of using the robot for positioning and moving parts - the use of small-sized and cost-effective robots in the production of small parts - simple laser beam feed system	- difficulties in machining large, heavy and less rigid parts - more difficult to program without a tool axis point (when programming a tool axis) - complexity of parts assembly
II	- low-profile, low-cost robots - it is possible to create production lines and integrate with other robots - a small footprint is needed for the workstation - easy to program with a tool axis point - machining of large, heavy or lightly rigid parts - reaching into the interior or under the workpiece (e.g., machine tool body)	- limited working space in size and reach of robot arms - possible collision of the beam delivery system with the object, it is necessary to simulate the path of the robot's movement with the laser and preliminary testing of working movements at lower speeds before starting production - potentially higher laser power losses due to multiple possible beam bends
III	- large working space - smooth motion - ease of programming - simple optical system with standard mirrors	- taking up a significant amount of production space - no machining possible underneath the part - more difficult to add additional 2-axis motion - complex integration with additional robot - higher cost
IV	- an increase in the number of machining operations per station - simplifying and shortening the time of the technological process - potential lower number of defective products	- high cost - reduced availability of equipment - advanced staff competences are needed - use of laser to process a limited number of materials (dependence of laser processing on the intended use of the machine tool)

The different types of robotized laser stations have their own capabilities and limitations (Oczoś K., 1988).

In robotized laser stations there are two ways of bringing the beam to the workpiece (Industry, 2017):

- flexible (telescopic) beam delivery systems use guiding mirrors that reflect the beam towards the focusing optical system, e.g., in CO_2 lasers,
- the fiber optic cable in the form of high-quality quartz fibres leads the radiation to the laser work head, e.g., in Nd:YAG lasers.

Advantages of laser workstation robotization (Oczoś K., 1988):

- capability to cut three-dimensional, asymmetrical objects which are difficult to produce by means of other methods,
- flexible production of various items, individually tailored to the customer's order on a unit, serial or mass scale,

- accuracy and repeatability of beam travel,
- possibility of designing a trajectory of motion,
- reduced preparation time before machining,
- easy testing and simulation based on the control program,
- easy changeover and adaptation to machining in the case of parts with small differences in the technological process,
- cost savings in the context of preparing special equipment (no special handles or the use of folding handles),
- accuracy and reliability of machining.

other:

- no chips,
- no coolants, lubricants or other operating fluids,
- material saving (e.g., optimization of sheet metal cutting).

Just in Time and Lean Manufacturing

Automation and robotization - in addition to the organization of human work - are the pillars of the modern approach that has spread in recent years in the context of manufacturing. It is supported by two concepts - **Just in Time** and **Lean Manufacturing** (Womack J. & Others, 2011).

The Just in Time concept is to produce and deliver finished products exactly when they are needed (there is a customer), components when they are needed in the assembly operation, individual parts when they are needed for the assembly of components, and raw materials when they are needed to start the process of machining components. JiT's goals are:

- no non-compliant components - 0% of deficiencies,
- proper quantity - no less (there will be too little), no more (you have to store it in a warehouse) than needed (exactly as much as you need),
- punctuality (on time) - not too early or too late.

Achieving the objectives of JiT is possible only when:

- modular products are designed, easy to manufacture, without unnecessary complexity,

- production is organized in a direct line, evenly loaded with production tasks, synchronized in time with other elements of the production system,
- robots and manipulators are used to automate multiple operations,
- losses are reduced, i.e., losses resulting from: excessive number of produced items (in reserve), defective products, waste of time (stoppages, waiting time, unexpected equipment repairs, searching for tools at badly organized workstations, unnecessary activities),
- application of total quality management - the aforementioned TQM applies to every department of the company, and employees actively participate in the creation of quality and the culture of production,
- transport is properly organized - both within the production halls and between external partners (suppliers, service providers, partners),
- production batch is reduced to a minimum - stocks of work in progress are reduced, there is no excessive storage of products in the vicinity of workstations.

The Lean Manufacturing concept, launched at Toyota Motor Company's post World War II facilities, is now the most effective way to increase productivity, reduce waste and shorten production cycles. According to (Womack J. & Others, 2011), 'lean production' results from the fact that the manufacturing process consumes a smaller number of all factors, compared with conventional mass production.

"Less" is referred to as:

- half of the human effort,
- half of the production space,
- half of the financial resources invested in instrumentation and tools,
- half of the time that engineers spend on developing a new product,
- half of the time spent on the implementation of the new product into production,
- half of the inventory kept,
- half of the existing number of defects and quality problems,

in order to increase the range of manufactured products.

The implementation of the LM concept requires changes in the following areas:

- **organizational**: training, staff knowledge, organization of work, maintaining high quality of work and work culture, ensuring the quality of products. The methods used are preventive in nature and allow to maintain the developed standards for a longer period of time. The following methods should be mentioned: 5S - used for workplace organization, FMEA - Failure Mode Effects Analysis - for identification of potential defects, downtime, making it possible to take action to prevent the occurrence of a problematic situation (FMEA is used in relation to: products, processes, environmental aspects, fleet management);
- **technical**: the organization of a direct production line is associated with the abandonment of very efficient, specialized machines in favor of the use of smaller machines, enriched with special equipment that increases their efficiency, allows to shorten changeover times, and at the same time is a way of flexible production, because of the ease of changing the assortment of products;
- **production**: minimizing machine changeover times (SMED - Single Minutes Exchange of Die - striving for machine changeover within a single digit), value stream mapping (VSM) and Poka-Yoke - to identify and avoid waste in production.

Concept Industry 4.0

Increasing flexibility, frequent changes in assortment and significant improvement in production efficiency are the basis for a modern approach to production. Contemporary market needs and changes in customer expectations force manufacturing plants to comprehensively rebuild their thinking about manufacturing. It can be said that the previous approach: the JiT system, the LM concept, is becoming insufficient for the currently observed, dynamic changes in the industry.

The term "Industry 4.0" derives from a certain consistency in the changes observed - the so-called "milestones" or "revolutions" - which were key stages in the development of industry in the world (Stadnicka D. & Others, (2017):

- **Industry 1.0** - this is the stage of **mechanization**. In the past, when the steam engine was invented and implemented, the world entered the era of industrialization. In industrial plants, workstations were mechanized and steam engines were commonly used. The small scale manufacturing of handmade goods was abandoned.

- **Industry 2.0** - this is the **electrification** phase. Along with electricity, the concept of linking individual workstations to production lines was put forward, and the opportunities offered by electricity led to the launch of large-scale manufacturing of products.
- **Industry 3.0** - this is the stage of **digitalization**. The birth and a widespread use of computers, with unprecedented data processing capabilities, have enabled businesses to grow further. Digitalization, increasingly efficient processing and control systems and modern software have made production machines efficient, precise, offering the possibility of combining them into flexible production systems, in which planning and control systems were an integral part. Digitalization has also enabled a higher degree of automation, which is a determinant of the level of technology development.
- **Industry 4.0** is an ongoing **integration** phase. The assumption of the new, revolutionary approach is to integrate various systems and create networks, connecting people and machines (personnel and all fixed assets). In addition, there is integration with previously independent stages or areas, e.g., with the partners with whom the company co-creates the so-called value chain. Now they are managed and optimized interdependently.

The Industry 4.0 concept was developed in 2011 in Germany. It involves transforming existing factories into self-steering and self-adaptive Smart Factories to create intelligent value chains. Nowadays, this concept has been implemented to the greatest extent in Germany and the USA, where even demonstration plants are being built to train manufacturers and managers.

The Industry 4.0 concept is based on a shift from manufacturing processes that were previously aimed at mass production of identical products to **mass production of individualized products**. This requires an extraordinary, flexible design of production lines in which individual technological processes can be carried out. This flexibility, on an unprecedented scale, is possible thanks to comprehensive computerization, automation and robotization of processes.

In order to be able to comprehensively manage and implement many individualized technological processes in a short period of time, the information flow is designed between the company components and the IT department (vertical) and between the machines and the production system (horizontal) (Muhlemann A.P. & Others, 2001).

Horizontal and vertical integration is shown in the diagram (Figure 5).

Figure 5. Horizontal (a) and vertical (b) flow of values

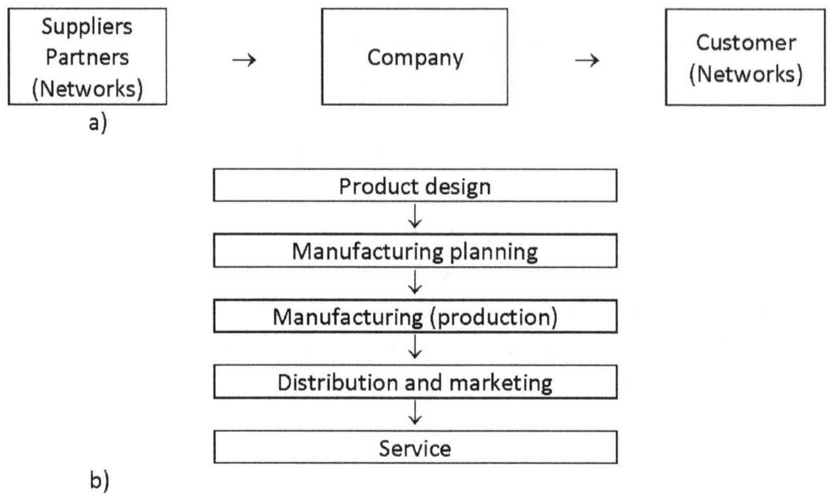

The intelligent production management systems used in the transformed plants are based on on-line communication. For this reason, changes are also required at the digitalization level. All areas of the company's activity are supported by intelligent decision making systems and broadly understood automation. In this concept one can distinguish areas of change concerning the functioning of enterprises, such as (Stadnicka D. & Others, (2017):

- software - a tool for integrating various functions of various devices and systems, typical for CIM (Computer Integrated Manufacturing), enabling the management and operation of the elements of the entire production cycle. It enables handling customer orders, issuing orders to complete materials, tools, workstations, enables automatic activation of the necessary sequence of machining and transport commands, switching off unnecessary commands, collecting data on the condition of machines, automatic switching off damaged modules, signalling failures, monitoring the quantity and quality of products, generating reports, etc.,
- cloud data collection and processing - transformed production management processes rely on the use of a huge amount of data generated by control systems. Previously, they were used mainly to monitor the state of technological processes; in I4.0 they enable prediction of behavior (methods of implementation, modifications of

machine tools, etc.), guaranteeing quality parameters of processes, production control in a global perspective. This makes it possible to analyze and generate decisions based on a huge amount of data,

- cyber security issues - in case of hacking attacks, takeover of know-how, change of functions and working parameters of devices to the detriment of the owner-creator,
  - the use of artificial intelligence, and a number of others, related to comprehensive computerization and data flow on an unprecedented scale. Data in a huge amount, the so-called Big Data - that is the patterns of production processes, trends (e.g., according to customer preferences, changes in production capacity, materials used, the need to adapt production strategies to economy, market, fashion, finance, legislation, etc.), text, image, video, stored, used in calculations, simulations, transmitted - are counted in petabytes. Data collected in the cloud should be supported by decision making systems based on elements of artificial intelligence (Stadnicka D. & Others, (2017). Then, the implementation of various technological operations and production management in the broad sense requires a minimum of time.

The key transformation factors to level 4.0 are:

- M2M communication - machine to machine,
- advanced methods of communicating information.

The integration of all the areas of a company makes it possible to respond quickly to market demand and involve the suppliers and customers in the processes of production, sales and distribution. SMART factories are to be intelligent, production lines will automatically modify operating parameters, devices will exchange information between each other, quickly and effectively modify product configuration. Production will be more flexible and manufacturers will be able to deliver more complex orders faster and cheaper than before.

A change in the way information is collected and processed is also necessary to achieve comprehensive integration.

It is fostered by:

- digitalization and integration of value chains - mobile devices, laptops, smartphones, tablets are used, data are stored and processed in the cloud, **Internet of Things** platforms are implemented,
- modernization of process management methods, customer access and other approaches in creating business models - modern technologies of location detection, advanced human-machine interfaces, advanced technologies for data security to detect fraud, hacking attacks, unauthorized interference in company systems are applied,
- digitalization of products and services - in the so-called augmented or virtual reality, analysis of data according to customer profiles (including the possibility of customer profiling, multi-level interactions with customers), advanced algorithms of data processing at the stage of: design, engineering calculations, simulation and optimization of technological processes; application of intelligent sensors to transmit in real time information about deviations in the production process (especially during processing, transport, storage) and about the properties of the products.

In view of the need to process a wide variety of orders from many customers in a short period of time, it is necessary to digitalize all processes within the company. The concept of Industry 4.0 assumes that such an approach is based on system integration and networking of information flows between people, but also between people and machines.

In order for the vision of Industry 4.0 to be realized, all elements of the production line should be compatible with each other. To satisfy the requirements, numerous producers can supply companies with both machine tools and other components, e.g., sorting equipment, conveyors, interchangeable tables with complete equipment, and process actuators with matching software (Industry, 2017*)*. Then, you can say that the supplier offers the entire factory (Figure 6).

The widespread use of the Internet and cloud computing is visible in SMART factories. In the so-called Cloud Computing, three solutions are offered to entrepreneurs:

- software provided by the Interent - Software as a Service (SaaS),
- platforms - computing, contacts with customers, suppliers - Platform as a Service (PaaS),
- infrastructure - Infrastructure as a Service (IaaS).

Figure 6. Integrated factory: a) on-line process supervision, b) process control, c) sorting of parts at the workstation, d) high bay warehouses

This gives certain benefits: no need to buy your own hardware and software, availability of solutions which are always up-to-date, using and paying only for the currently needed resources.

Industry 4.0 has so far been implemented in companies in sectors such as:

- motoring,
- aerospace, defence, security and security industries,
- chemical industry,
- electronic industry,
- metal, construction industry,
- transport and logistics,
- medicine,

in spite of the wide range of applications and the diversity of the manufactured constructions.

The Future of Lasers in Industry 4.0

The assumption of Industry 4.0 is to create a logical sequence of technological operations enabling immediate changeover and adjustment of processing parameters to the order.

Due to the achievements in laser technology it can be said that lasers are the technology of the future for Industry 4.0. A great variety of processing options and ways of using laser energy offers a wide range of possibilities for building integrated production lines.

Laser heads placed in machining sockets for cutting enable combined machining, e.g., punching, drilling, welding, etc., with simultaneous reduction of production time (Skoczylas A., 2013).Lasers often replace conventional workstations, e.g., for cutting, engraving, hardening, etc. Additionally, laser **measuring systems**, various types of sensors and code readers complement machining operations with identification of product features, measurements or marking (Figure 7).

Figure 7. Product marking: letters, digits, QR codes, barcode - according to the customer's order

An additional advantage of laser technology is the ease of changing **processing parameters** (loading the appropriate file), cleanliness at workstations, safety (acceptable risk), benefits in relation to the postulates of waste-free or low-waste production, energy-saving, without chemicals, the possibility of processing products of virtually unlimited shapes (3D printing technique) or sizes (Figure 8).

Figure 8. Products made to order in 3D laser techniques

From the perspective of technological processing, lasers have become popular as a result of the advantages associated with the process: reproducibility of settings, potential for automation, controllability, durability of the laser beam as a tool. In addition, they are a source of moderate noise, small quantities of harmful agents (provided that adequate protection is in place) as well as waste and material losses. They do not show any defects typical of machining. Laser causes slight self-stress in materials and makes it possible to cut out complicated shapes of elements, and objects usually do not require finishing (Zaborski S., Stechnij T., 2011; Skoczylas A., 2013). There are also well-known technological solutions that involve, e.g., the production of laser punching machines that can cooperate with robots, and are equipped with tilt heads or systems enabling beam division and simultaneous multi-station processing (Trzepieciński T., 2012; Plastics, 2009).

Robotization is common in the case of laser technology. The devices at the workstations not only perform tasks (welding, marking), but are also equipped with sensors, control systems and machine learning algorithms (artificial intelligence). This enables not only machining but also self-monitoring as

Figure 9. Cut and stamped product

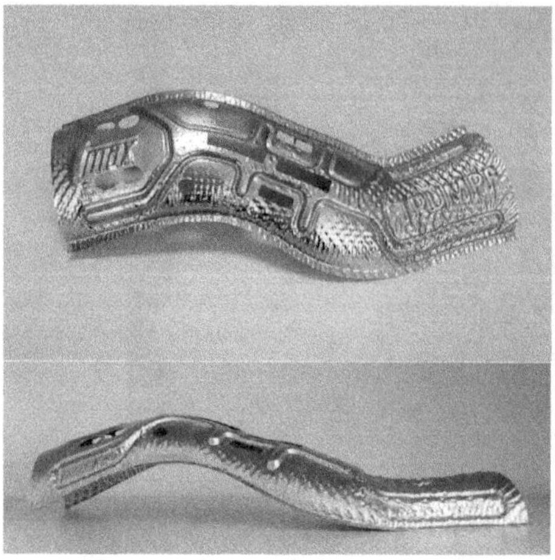

Figure 10. Cut, bent, welded and laser engraved product

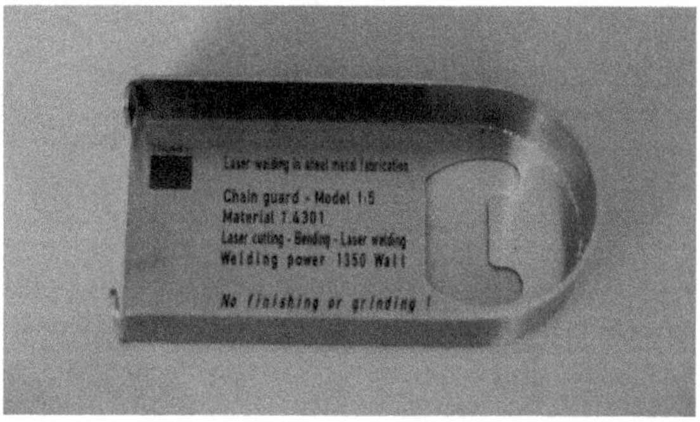

Figure 11. Laser die-cut and bent on two separate devices

well as breakdown and downtime prevention. Robotized laser devices are additionally equipped with ambient detection sensors. They prevent damage to the workplace and reduce the risks to people arising from quiet and less noticeable machine operation. Sensors are mounted e.g., in joints, they detect or prevent collisions, they can slow down or stop moving parts or change their trajectory to a safe one.

Laser technology brings enormous manufacturing potential to the industry strategy 4.0. Many laser manufacturers have implemented new forms of cooperation between customers and production managers. Laser is not just an actuator. Devices equipped with laser heads make it possible, thanks to additional systems, to perform combined operations, i.e., combining modern laser processing with operations typical of conventional machine tools, such as punching, drilling, threading, welding, rolling, while significantly reducing the production time (Laser, 2020).

Figure 12. Example of a part made on a laser punching machine with additional tooling for pressed parts - TRUMPF Combi type machine

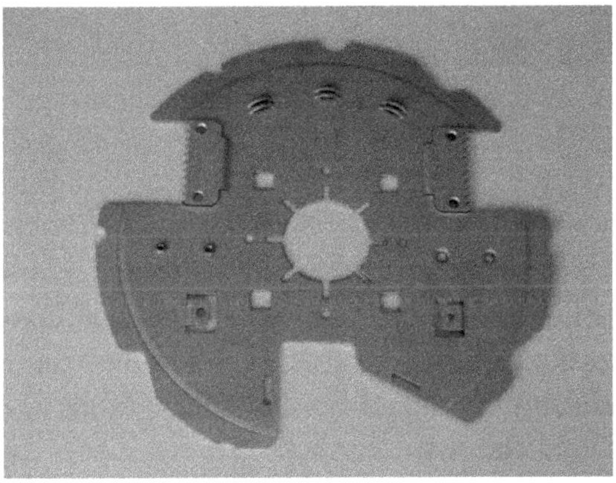

Figure 13. Examples of laser cut products for 3D profiles - TruLaser Tube 5000 fiber

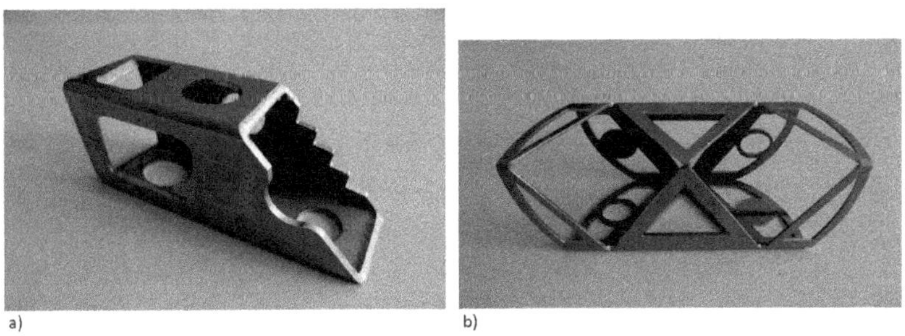

a) b)

Technological operations are planned and carried out with the use of additional equipment for sorting parts, robotic feeders and organization of temporary storage zones for parts. These machines require advanced optical systems, status sensors and control systems (Industry, 2017). They enable the transfer of data about the state of the machine by Cloud via the Internet. In this case, the requirements for communication between machines and people are met.

Figure 14. Sheet metal prepared for cutting, laser for cutting long pipes

Such a solution is used in flexible lines for cutting, marking, laser welding.

Sometimes, in the process of building a logical sequence of technological operations, lasers are connected with other devices, e.g., bending machines. Then, the produced element does not leave the technological line which is an automated and robotized production line. Another solution used in laser applications is building multihead devices (multi-head systems) which can be found in SLM applications.

Lasers for 3D printing are also used in the production line. They are characterized by specific features, such as:

- it is easy to personalize production - beginning a new production cycle of the ordered product starts with entering the program, which is not time-consuming,
- can be produced on request - ideal for Just in Time,
- easy set-up - one machine can perform commands from different integrated and computer-controlled production lines,

- production of lighter products (environmental postulate of material savings),
- production of complete components, mechanisms, lack of assembling operations,
- use of materials that are difficult to process (metal powders, metals with ceramics).

The change of laser working parameters gives the possibility of processing regardless of the material characteristics. It is possible to manufacture a variety of products to customer order, on a unit scale (Figure 15 and 16).

Figure 15. Products made with a TRUMPF 2D CO_2 laser (a - structural steel, b – stainless steel)

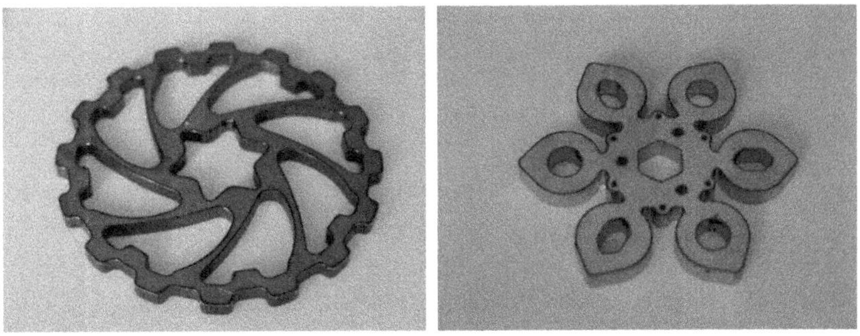

Figure 16. Laser cutting of large and thick products, and very small products

An additional issue is to define the customer's order in such a way that the features of a single product can be translated into the language of machines (Figure 17).

Figure 17. Real-time monitoring of production status

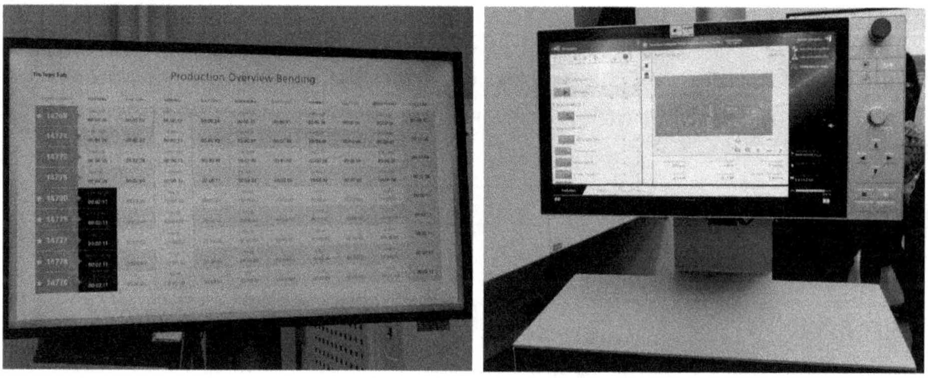

Many laser manufacturers provide workstations with software to connect databases with device drivers. This is an advantage due to the much shorter lead time compared to conventional technologies, ease of cleaning and safety at workstations. Lasers can also be used to mark products as a result of the need to quickly orientate the objects in the production line. If a product is to have individual characteristics, it should be easily distinguishable from other products. Lasers enable fast coding for traceability, order number and quality control (Industry, 2017). Labelling also enables the use of easy-to-install components, like PC, electronics, beam source and optics.

DIRECTIONS FOR THE IMPROVEMENT OF LASER TECHNOLOGY

Producers are expected not only to attain high quality and efficiency but also to reduce costs and, most importantly, to be able to respond quickly and effectively to the needs of the international market. Advances in technology lead to a strategy called agile manufacturing.

Agile manufacturing requires organizing the flow of information within the company to comprehensively assess the manufacturing capacity of many different product variants. Sometimes these variants exist only in virtual reality that can be created thanks to advanced computer techniques.

Lasers used in modern production are, one could say, perfectly in line with these tendencies. This is facilitated by the variety of laser designs and applications. Additionally, it is possible to control the devices by means of computers as well as to robotize and automate them. The advantages of lasers are their functional properties when used as manufacturing tools.

Lasers are becoming more efficient as well as durable tools, and it is also possible to eliminate the problems encountered in conventional machining. Examples are titanium and its alloys, magnesium and magnesium alloys.

The properties of titanium alloys mean that cutting machines must have high dynamic stiffness and vibration damping capability. These conditions must be met to ensure the accuracy of the dimensions and shape of the surface to be machined.

Machine tools are equipped with active vibration damping systems, massive beds, stiffer columns, wide guides and drives adapted to heavy loads. In addition, cutting titanium alloys requires lowering the cutting speed and increasing the torque. The reason for this is the use of tools whose smaller diameters provide acceptable durability. Both conditions allow to maintain proper productivity of the production process. These requirements make such machines inexpensive.

In the case of magnesium, the problem that should be taken into account is the intense heating of the material and the formation of finely atomized chips in the case of grinding.

During grinding, particles can agglutinate the grinding wheels, which in turn can lead to rapid wear or cracking of the wheel, but also pose an explosion hazard.

There is a significant fire hazard during the processing of magnesium alloys. Machines with ventilation devices, fire and explosion protection systems are used on machining stations. The swarf is transported from the cutting zone to the work area where it is dried and agglomerated. Modern machining centres also have aerosol filtration systems, temperature, pressure, light intensity and argon extinguishing systems. In some cases, the work of a human being, consisting in loading and unloading a machine tool, is done by a robot. Even more reactive are beryllium alloys, which require special care in removing the smallest particles in order to prevent ignition and explosion (Oczoś K., Kawalec A., 2012).

These problems do not occur at all, or occur on a much smaller scale, during laser processing.

Not everything has yet been achieved in laser technology. For example, for 3D printing, implementation studies are being carried out on the possibility of extending the range of materials used, printing many different products simultaneously with one device, increasing availability in terms of part dimensions (still limited by the container with hardened material), improving the durability and quality of printed products.

Lasers offer new possibilities in material bonding technology. It can be said that in a situation where it is not justified to give up adhesives (as well as the noxiousness associated with the traditional method of gluing), preparation with a laser surface is the first stage of improving manufacturing methods. The next step is to combine different materials with each other, of different thickness, shape, in structures difficult to access, without additional equipment, which is possible by means of sport welding. It is possible to combine thin-walled elements made of different materials (Figure 18).

Figure 18. Combination of different materials from SPI Lasers: of stainless, aluminium, copper and brass to stainless steel

Instead of spot welding, TRUMPF offers spot welding by melting. A laser beam with appropriately selected parameters melts the surface layer of the material that cools down and hardens for a longer period of time, and then melts the surface layer of the material that hardens faster. Elements are brought close to each other, slightly pressed, to achieve a permanent connection (Figure 19).

Figure 19. Example of two laser-connected parts made of different types of materials: plastic and metal. The connection point has been melted; the elements are close to each other and then pressed together

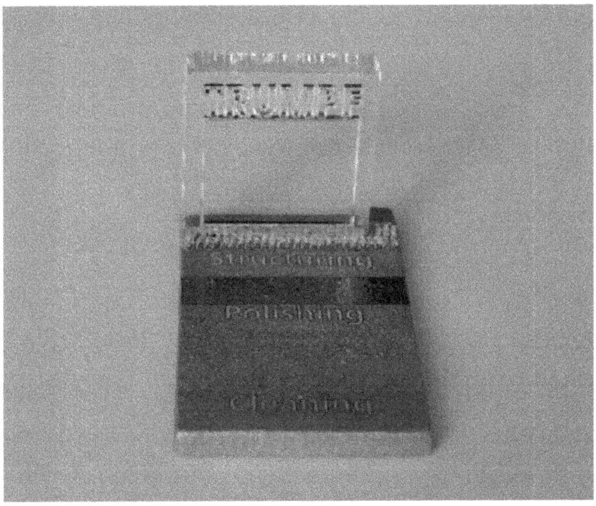

CONCLUSION

The considerations presented in this chapter lead to the conclusion that lasers will become more widespread and popular in the production of many products in the coming years. Such conclusions may be based on the changes observed in production management - greater emphasis is put on ecological activities, there is a mass production of individual products, as well as intensive modernization of equipment. Preference is given to devices that can be used to create production lines, which are easy to control or modify and are convenient in robotization or automation. The expected features can be successfully attributed to laser devices, whose production capabilities are very wide, and their modification and control can be done by means of computer software. The progress visible in the environmental management sciences, the construction of machine tools and manufacturing systems gives reasons for hope that adhesive bonding technology, which is the leading feature of this monograph, will also undergo a technological change that will benefit human health and the quality of the environment. The symptoms of this change are already noticeable, as demonstrated in the given examples.

REFERENCES

Badida, M., Ladomersky, J., Kralikova, R., Moravec, M., & Liptai, P. (2014). *Environmentalne inzinierstvo* [Environmental engineering]. Kosice: ELFA, s.r.o.

Bernaciak, A., & Gaczek, W. M. (2001). Ekonomiczne aspekty ochrony środowiska [Economic aspects of environmental protection]. Poznań: Wyd. Akademii Ekonomicznej.

Cato, W., & Mobley, R. (2002). Computer-Managed Maintenance Systems. Butterworth-Heinemann.

Groover, M. (2008). *Automation, Production Systems, and Computer-Integrated Manufactirung*. Pearson Education Inc.

Honczarenko, J. (2000). *Elastyczna automatyzacja wytwarzania. Obrabiarki i systemy obróbkowe* [Flexible manufacturing automation. Machine tools and machining systems]. Warszawa, WNT.

Industry 4.0: Benefit From Opportunities. Today. Tomorrow. (2012). *Laser Technik Journal, 5*, 12-14.

ISO 14001:2015 *Environmental, management systems – Requirements with guidance for use*

Laser machinne with additional functions. (n.d.). TruMatic 1000 fiber. Materials TRUMPF Poland.

Lis, S., Santarek, K., & Strzelczak, S. (1994). Organizacja elastycznych systemów produkcyjnych [Organization of flexible production systems]. Warszawa: PWN.

Muhlemann, A. P., Oackland, J. S., & Lockyer, K. G. (2001). *Production and operactions management*. Warszawa: PWN.

Oczoś, K. (1988). Kształtowanie materiałów skoncentrowanymi strumieniami energii [Shaping materials with concentrated energy streams]. Rzeszów: Wydawnictwo Politechniki Rzeszowskiej.

Oczoś, K., & Kawalec, A. (2012). Kształtowanie metali lekkich [Light metal shaping]. Warszawa: PWN.

Plastics. (2009): *Some words about machining of plastics*. Retrieved from https://www.cadblog.pl/archiwum/CADblogpl_nr_3(4)_2009_LQ_v2.pdf

Skoczylas, A. (2013). Analiza porównawcza procesu cięcia wiązką laserową i strumieniem wodno-ściernym [Comparative analysis of the laser beam and abrasive waterjet cutting process]. *Postępy Nauki i Techniki, 8*, 121–128.

Stadnicka, D., Zielecki, W., & Sęp, J. (2017). *Koncepcja przemysł 4.0 – ocena możliwości wdrożenia na przykładzie wybranego przedsiębiorstwa* [Industry 4.0 concept - evaluation of implementation possibilities on the example of a selected enterprise]. Konferencja Innowacje w Zarządzaniu i Inżynierii Produkcji, Zakopane 2017. Retrieved from http://www.ptzp.org.pl/files/konferencje/kzz/artyk_pdf_2017/T1/t1_472.pdf

Trzepieciński, T. (2012). Trendy rozwojowe maszyn i technik stosowanych w technologiach cięcia blach [Development trends of machines and techniques used in sheet metal cutting technologies]. *Inżynieria Maszyn, 17*(3), 94–106.

Womack, J., Jones, D., & Ross, D. (2011). *Machine That Changed the World*. New York: Simon&Shuster.

Zaborski, S., & Stechnij, T. (2011). Laserowe i plazmowe cięcie blach ze stali niestopowych i kwasoodpornych [Laser and plasma cutting of sheets made of non-alloy and acid-resistant steel]. *Inżynieria Maszyn, 16*(4), 109–116.

Related Readings

To continue IGI Global's long-standing tradition of advancing innovation through emerging research, please find below a compiled list of recommended IGI Global book chapters and journal articles in the areas of public administration, democracy, and social movements. These related readings will provide additional information and guidance to further enrich your knowledge and assist you with your own research.

Abdel-Aal, H. A. (2015). The Structure of Ventral Scale Textures in Snakes in Comparison to Texturing of Deterministic Tribological Surfaces. In R. Tyagi & J. Davim (Eds.), *Processing Techniques and Tribological Behavior of Composite Materials* (pp. 268–315). Hershey, PA: IGI Global. doi:10.4018/978-1-4666-7530-8.ch010

Abdulrahman, K. O., Akinlabi, E. T., & Mahamood, R. M. (2019). Additive Manufacturing: Laser Metal Deposition and Effect of Preheating on Properties of Deposited Ti-4822-4 Alloy. In K. Kumar, D. Zindani, & J. Davim (Eds.), *Additive Manufacturing Technologies From an Optimization Perspective* (pp. 165–183). Hershey, PA: IGI Global. doi:10.4018/978-1-5225-9167-2.ch008

Adebiyi, I. D., Popoola, P. A., & Pityana, S. (2016). Mitigation of Wear Damage by Laser Surface Alloying Technique. In E. Akinlabi, R. Mahamood, & S. Akinlabi (Eds.), *Advanced Manufacturing Techniques Using Laser Material Processing* (pp. 172–196). Hershey, PA: IGI Global. doi:10.4018/978-1-5225-0329-3.ch007

Aida, A. (2020). Fluid Matters: The Water Footprint of Building Materials. In B. Brownell (Ed.), *Examining the Environmental Impacts of Materials and Buildings* (pp. 160–186). Hershey, PA: IGI Global. doi:10.4018/978-1-7998-2426-8.ch006

Ali, S., & Murari, V. (2019). Miniaturization of Test Specimen for Composites. In M. Uthayakumar, S. Raj, T. Ko, S. Kumaran, & J. Davim (Eds.), *Handbook of Research on Green Engineering Techniques for Modern Manufacturing* (pp. 49–76). Hershey, PA: IGI Global. doi:10.4018/978-1-5225-5445-5.ch004

Ali, S. F., Malik, F. M., Kececi, E. F., & Bal, B. (2019). Optimization of Additive Manufacturing for Layer Sticking and Dimensional Accuracy. In K. Kumar, D. Zindani, & J. Davim (Eds.), *Additive Manufacturing Technologies From an Optimization Perspective* (pp. 185–198). Hershey, PA: IGI Global. doi:10.4018/978-1-5225-9167-2.ch009

Ameddah, H., & Mazouz, H. (2019). 3D Printing Analysis by Powder Bed Printer (PBP) of a Thoracic Aorta Under Simufact Additive. In K. Kumar, D. Zindani, & J. Davim (Eds.), *Additive Manufacturing Technologies From an Optimization Perspective* (pp. 102–118). Hershey, PA: IGI Global. doi:10.4018/978-1-5225-9167-2.ch005

Amuda, M. O., & Akinlabi, E. T. (2016). Trend and Development in Laser Surface Modification for Enhanced Materials Properties. In E. Akinlabi, R. Mahamood, & S. Akinlabi (Eds.), *Advanced Manufacturing Techniques Using Laser Material Processing* (pp. 92–120). Hershey, PA: IGI Global. doi:10.4018/978-1-5225-0329-3.ch005

Anand, G., & Chattopadhyay, P. P. (2016). Computational Design of Microstructure: An Overview. In S. Datta & J. Davim (Eds.), *Computational Approaches to Materials Design: Theoretical and Practical Aspects* (pp. 61–100). Hershey, PA: IGI Global. doi:10.4018/978-1-5225-0290-6.ch004

Andriotis, A. N., & Menon, M. (2016). Successive Spin-Correlated Local Processes Underlying the Magnetism in Diluted Magnetic Semiconductors and Related Magnetic Materials. In S. Datta & J. Davim (Eds.), *Computational Approaches to Materials Design: Theoretical and Practical Aspects* (pp. 13–27). Hershey, PA: IGI Global. doi:10.4018/978-1-5225-0290-6.ch002

Arumugaprabu, V. R., D. J., & R., P. (2019). Effective Utilization of Industrial Wastes for Preparing Polymer Matrix Composites: Usage of Industrial Wastes. In M. Uthayakumar, S. Raj, T. Ko, S. Kumaran, & J. Davim (Eds.), Handbook of Research on Green Engineering Techniques for Modern Manufacturing (pp. 250-261). Hershey, PA: IGI Global. doi:10.4018/978-1-5225-5445-5.ch013

Ashrafiyan, O., Saremi, M., Pakseresht, A., & Ghasali, E. (2018). Oxidation-Protective Coatings for Carbon-Carbon Composites. In A. Pakseresht (Ed.), *Production, Properties, and Applications of High Temperature Coatings* (pp. 429–446). Hershey, PA: IGI Global. doi:10.4018/978-1-5225-4194-3.ch016

Assefa, G. (2020). Life-Cycle Assessment-Based Environmental Performance Targets for Buildings: What Is Next? In B. Brownell (Ed.), *Examining the Environmental Impacts of Materials and Buildings* (pp. 188–216). Hershey, PA: IGI Global. doi:10.4018/978-1-7998-2426-8.ch007

Azadi, M. (2018). Deposition of TiN, TiC, and DLC Coatings by PACVD. In A. Pakseresht (Ed.), *Production, Properties, and Applications of High Temperature Coatings* (pp. 381–402). Hershey, PA: IGI Global. doi:10.4018/978-1-5225-4194-3.ch014

B. J. M. (2018). Studies on Metal Oxide Nanoparticle Doped PVP Polymer Nanocomposites. In R. Somashekar, & T. Urs G. (Eds.), Emergent Research on Polymeric and Composite Materials (pp. 209-224). Hershey, PA: IGI Global. doi:10.4018/978-1-5225-3023-7.ch009

Balachandran, P. V., & Rondinelli, J. M. (2016). Informatics-Based Approaches for Accelerated Discovery of Functional Materials. In S. Datta & J. Davim (Eds.), *Computational Approaches to Materials Design: Theoretical and Practical Aspects* (pp. 192–223). Hershey, PA: IGI Global. doi:10.4018/978-1-5225-0290-6.ch007

Balla, V. K., Das, M., Datta, S., & Kundu, B. (2015). Articulating Biomaterials: Surface Engineering, Tribology, and Biocompatibility. In R. Tyagi & J. Davim (Eds.), *Processing Techniques and Tribological Behavior of Composite Materials* (pp. 218–267). Hershey, PA: IGI Global. doi:10.4018/978-1-4666-7530-8.ch009

Barua, R., Datta, S., Roychowdhury, A., & Datta, P. (2019). Importance of 3D Printing Technology in Medical Fields. In K. Kumar, D. Zindani, & J. Davim (Eds.), *Additive Manufacturing Technologies From an Optimization Perspective* (pp. 21–40). Hershey, PA: IGI Global. doi:10.4018/978-1-5225-9167-2.ch002

Bathrinath, S., Saranyadevi, S., Thirumalai Kumaran, S., & Saravanasankar, S. (2019). PageRank Algorithm-Based Recommender System Using Uniformly Average Rating Matrix. In M. Uthayakumar, S. Raj, T. Ko, S. Kumaran, & J. Davim (Eds.), *Handbook of Research on Green Engineering Techniques for Modern Manufacturing* (pp. 99–112). Hershey, PA: IGI Global. doi:10.4018/978-1-5225-5445-5.ch006

Bhuyan, D. (2019). Design of Prosthetic Heart Valve and Application of Additive Manufacturing. In K. Kumar, D. Zindani, & J. Davim (Eds.), *Additive Manufacturing Technologies From an Optimization Perspective* (pp. 120–133). Hershey, PA: IGI Global. doi:10.4018/978-1-5225-9167-2.ch006

Bodratti, A. M., Cheng, C., & Alexandridis, P. (2015). Bridging Product Design with Materials Properties and Processing: An Innovative Capstone Course. In H. Lim (Ed.), *Handbook of Research on Recent Developments in Materials Science and Corrosion Engineering Education* (pp. 1–20). Hershey, PA: IGI Global. doi:10.4018/978-1-4666-8183-5.ch001

Bolleddu, V., Racherla, V., & Bandyopadhyay, P. P. (2018). Microstructural and Tribological Characteristics of Air Plasma Sprayed Alumina-Titania Coatings. In A. Pakseresht (Ed.), *Production, Properties, and Applications of High Temperature Coatings* (pp. 268–298). Hershey, PA: IGI Global. doi:10.4018/978-1-5225-4194-3.ch011

Boonkerd, K. (2017). Development and Modification of Natural Rubber for Advanced Application. In T. Kobayashi (Ed.), *Applied Environmental Materials Science for Sustainability* (pp. 44–76). Hershey, PA: IGI Global. doi:10.4018/978-1-5225-1971-3.ch003

Boonmahitthisud, A. (2017). Natural Rubber and Rubber Blend Nanocomposites: Reinforcement of Natural Rubber with Polymer-Encapsulated Inorganic Nanohybrid Particles. In T. Kobayashi (Ed.), *Applied Environmental Materials Science for Sustainability* (pp. 77–105). Hershey, PA: IGI Global. doi:10.4018/978-1-5225-1971-3.ch004

Boschetto, A., & Bottini, L. (2014). Surface Characterization in Fused Deposition Modeling. In L. Santo & J. Davim (Eds.), *Surface Engineering Techniques and Applications: Research Advancements* (pp. 249–280). Hershey, PA: IGI Global. doi:10.4018/978-1-4666-5141-8.ch008

Brownell, B. E. (2019). Determining Architecture's Footprint: Preliminary Methods for Measuring the True Environmental Impact of Buildings. In G. Koç & B. Christiansen (Eds.), *Reusable and Sustainable Building Materials in Modern Architecture* (pp. 28–59). Hershey, PA: IGI Global. doi:10.4018/978-1-5225-6995-4.ch002

Carrillo, K. L., & Kobayashi, T. (2017). Natural Material Source of Bagasse Cellulose and Their Application to Hydrogel Films. In T. Kobayashi (Ed.), *Applied Environmental Materials Science for Sustainability* (pp. 19–43). Hershey, PA: IGI Global. doi:10.4018/978-1-5225-1971-3.ch002

Chakraborti, N. (2016). Data-Driven Bi-Objective Genetic Algorithms EvoNN and BioGP and Their Applications in Metallurgical and Materials Domain. In S. Datta & J. Davim (Eds.), *Computational Approaches to Materials Design: Theoretical and Practical Aspects* (pp. 346–368). Hershey, PA: IGI Global. doi:10.4018/978-1-5225-0290-6.ch012

Chen, Z., & Wu, C. (2016). Ab Initio-Based Stochastic Simulations of Kinetic Processes on Surfaces. In S. Datta & J. Davim (Eds.), *Computational Approaches to Materials Design: Theoretical and Practical Aspects* (pp. 28–60). Hershey, PA: IGI Global. doi:10.4018/978-1-5225-0290-6.ch003

Choudhury, D., Roy, T., & Krupka, I. (2015). Surface Modifications and Tribological Effect in Orthopedics Implants. In R. Tyagi & J. Davim (Eds.), *Processing Techniques and Tribological Behavior of Composite Materials* (pp. 193–217). Hershey, PA: IGI Global. doi:10.4018/978-1-4666-7530-8.ch008

Cruse, A. (2020). Improving the Weather: On Architectural Comforts and Climates. In B. Brownell (Ed.), *Examining the Environmental Impacts of Materials and Buildings* (pp. 251–281). Hershey, PA: IGI Global. doi:10.4018/978-1-7998-2426-8.ch009

Cui, X., Zeng, S., Li, Z., Zheng, Q., Yu, X., & Han, B. (2018). Advanced Composites for Civil Engineering Infrastructures. In K. Kumar & J. Davim (Eds.), *Composites and Advanced Materials for Industrial Applications* (pp. 212–248). Hershey, PA: IGI Global. doi:10.4018/978-1-5225-5216-1.ch010

Dahmen, J., von Bergmann, J., & Das, M. (2020). Teardown Index: Emissions of Single-Family Homes in Vancouver. In B. Brownell (Ed.), *Examining the Environmental Impacts of Materials and Buildings* (pp. 64–101). Hershey, PA: IGI Global. doi:10.4018/978-1-7998-2426-8.ch003

Datta, S., Roy, B., & Davim, J. P. (2016). Computational Materials Design: Different Concepts and Aspects. In S. Datta & J. Davim (Eds.), *Computational Approaches to Materials Design: Theoretical and Practical Aspects* (pp. 1–12). Hershey, PA: IGI Global. doi:10.4018/978-1-5225-0290-6.ch001

Daver, F., & Hadgraft, R. (2015). Linking Materials Science and Engineering Curriculum to Design and Manufacturing Challenges of the Automotive Industry. In H. Lim (Ed.), *Handbook of Research on Recent Developments in Materials Science and Corrosion Engineering Education* (pp. 46–66). Hershey, PA: IGI Global. doi:10.4018/978-1-4666-8183-5.ch003

Dey, P., Bible, J., Dey, S., & Datta, S. (2016). Applications of Feature Selection and Regression Techniques in Materials Design: A Tutorial. In S. Datta & J. Davim (Eds.), *Computational Approaches to Materials Design: Theoretical and Practical Aspects* (pp. 224–251). Hershey, PA: IGI Global. doi:10.4018/978-1-5225-0290-6.ch008

Di Girolamo, G. (2018). Current Challenges and Future Perspectives in the Field of Thermal Barrier Coatings. In A. Pakseresht (Ed.), *Production, Properties, and Applications of High Temperature Coatings* (pp. 25–59). Hershey, PA: IGI Global. doi:10.4018/978-1-5225-4194-3.ch002

Dixit, A. (2018). Application of Silica-Gel-Reinforced Aluminium Composite on the Piston of Internal Combustion Engine: Comparative Study of Silica-Gel-Reinforced Aluminium Composite Piston With Aluminium Alloy Piston. In K. Kumar & J. Davim (Eds.), *Composites and Advanced Materials for Industrial Applications* (pp. 63–98). Hershey, PA: IGI Global. doi:10.4018/978-1-5225-5216-1.ch004

Elsayed, A. M., Dakkama, H. J., Mahmoud, S., Al-Dadah, R., & Kaialy, W. (2017). Sustainable Cooling Research Using Activated Carbon Adsorbents and Their Environmental Impact. In T. Kobayashi (Ed.), *Applied Environmental Materials Science for Sustainability* (pp. 186–221). Hershey, PA: IGI Global. doi:10.4018/978-1-5225-1971-3.ch009

Equbal, A., Equbal, M. A., Equbal, M. I., & Sood, A. K. (2019). Multi-Criterion Decision Method for Roughness Optimization of Fused Deposition Modelled Parts. In K. Kumar, D. Zindani, & J. Davim (Eds.), *Additive Manufacturing Technologies From an Optimization Perspective* (pp. 235–262). Hershey, PA: IGI Global. doi:10.4018/978-1-5225-9167-2.ch012

Erinosho, M. F., Akinlabi, E. T., & Pityana, S. (2016). Enhancement of Surface Integrity of Titanium Alloy with Copper by Means of Laser Metal Deposition Process. In E. Akinlabi, R. Mahamood, & S. Akinlabi (Eds.), *Advanced Manufacturing Techniques Using Laser Material Processing* (pp. 60–91). Hershey, PA: IGI Global. doi:10.4018/978-1-5225-0329-3.ch004

Faisal, N., & Kumar, K. (2019). Recycling and Reuse of Building Materials From Construction and Demolition: An Environmental Evaluation for Sustainable Growth. In G. Koç & B. Christiansen (Eds.), *Reusable and Sustainable Building Materials in Modern Architecture* (pp. 60–79). Hershey, PA: IGI Global. doi:10.4018/978-1-5225-6995-4.ch003

Fatoba, O. S., Popoola, A. P., Farotade, G. A., & Pityana, S. L. (2016). Computational Dynamics of Laser Alloyed Metallic Materials for Improved Corrosion Performance: Computational Dynamics of Laser Alloyed Metallic Materials. In E. Akinlabi, R. Mahamood, & S. Akinlabi (Eds.), *Advanced Manufacturing Techniques Using Laser Material Processing* (pp. 197–235). Hershey, PA: IGI Global. doi:10.4018/978-1-5225-0329-3.ch008

Figueira, R. B. (2018). Hybrid Sol-Gel Coatings: Erosion-Corrosion Protection. In A. Pakseresht (Ed.), *Production, Properties, and Applications of High Temperature Coatings* (pp. 334–380). Hershey, PA: IGI Global. doi:10.4018/978-1-5225-4194-3.ch013

Ganguly, S., & Datta, S. (2016). Imprecise Knowledge and Fuzzy Modeling in Materials Domain. In S. Datta & J. Davim (Eds.), *Computational Approaches to Materials Design: Theoretical and Practical Aspects* (pp. 252–266). Hershey, PA: IGI Global. doi:10.4018/978-1-5225-0290-6.ch009

Godi, A., & De Chiffre, L. (2014). Functional Surfaces in Mechanical Systems: Classification, Fabrication, and Characterisation. In L. Santo & J. Davim (Eds.), *Surface Engineering Techniques and Applications: Research Advancements* (pp. 1–44). Hershey, PA: IGI Global. doi:10.4018/978-1-4666-5141-8.ch001

Gupta, A. K., Dey, A., & Mukhopadhyay, A. K. (2016). Micromechanical and Finite Element Modeling for Composites. In S. Datta & J. Davim (Eds.), *Computational Approaches to Materials Design: Theoretical and Practical Aspects* (pp. 101–162). Hershey, PA: IGI Global. doi:10.4018/978-1-5225-0290-6.ch005

Guzmán, A. M., Maldonado, N. G., & Affranchino, G. (2019). Comprehensive Evaluation for Mortars and Concretes Incorporating Wastes. In G. Koç & B. Christiansen (Eds.), *Reusable and Sustainable Building Materials in Modern Architecture* (pp. 108–136). Hershey, PA: IGI Global. doi:10.4018/978-1-5225-6995-4.ch006

H, S., & Alabur, M. (2018). Structure-Property Relations in Polymers and Polymer Blends Using X-Ray Diffraction Technique. In R. Somashekar, & T. Urs G. (Eds.), Emergent Research on Polymeric and Composite Materials (pp. 122-138). Hershey, PA: IGI Global. doi:10.4018/978-1-5225-3023-7.ch005

Hassan, A., Alotaibi, J. G., Shalwan, A., & Yousif, B. F. (2015). Simulation of Fragmentation Technique Using ANSYS Software. In R. Tyagi & J. Davim (Eds.), *Processing Techniques and Tribological Behavior of Composite Materials* (pp. 341–372). Hershey, PA: IGI Global. doi:10.4018/978-1-4666-7530-8.ch012

Hassanzadeh, M., Saremi, M., Valefi, Z., & Pakseresht, A. H. (2018). Investigation on Improved-Durability Thermal Barrier Coatings: An Overview of Nanostructured, Multilayered, and Self-Healing TBCs. In A. Pakseresht (Ed.), *Production, Properties, and Applications of High Temperature Coatings* (pp. 60–78). Hershey, PA: IGI Global. doi:10.4018/978-1-5225-4194-3.ch003

Henao, J., & Sotelo, O. (2018). Surface Engineering at High Temperature: Thermal Cycling and Corrosion Resistance. In A. Pakseresht (Ed.), *Production, Properties, and Applications of High Temperature Coatings* (pp. 131–159). Hershey, PA: IGI Global. doi:10.4018/978-1-5225-4194-3.ch006

Hu, M. (2020). The Embodied Impact of Existing Building Stock. In B. Brownell (Ed.), *Examining the Environmental Impacts of Materials and Buildings* (pp. 1–31). Hershey, PA: IGI Global. doi:10.4018/978-1-7998-2426-8.ch001

Huyen, P. T. (2017). Clay Minerals Converted to Porous Materials and Their Application: Challenge and Perspective. In T. Kobayashi (Ed.), *Applied Environmental Materials Science for Sustainability* (pp. 141–164). Hershey, PA: IGI Global. doi:10.4018/978-1-5225-1971-3.ch007

Ida, A. (2019). Energetic Forms of Matter. In G. Koç & B. Christiansen (Eds.), *Reusable and Sustainable Building Materials in Modern Architecture* (pp. 137–165). Hershey, PA: IGI Global. doi:10.4018/978-1-5225-6995-4.ch007

Ikematsu, S., Tada, I., & Nagasaki, Y. (2017). Analysis of Lipids Produced by Microalgae Isolated from the Area around Okinawa, Japan. In T. Kobayashi (Ed.), *Applied Environmental Materials Science for Sustainability* (pp. 222–233). Hershey, PA: IGI Global. doi:10.4018/978-1-5225-1971-3.ch010

Ismail, M., Raja, P. B., & Salawu, A. A. (2015). Developing Deeper Understanding of Green Inhibitors for Corrosion of Reinforcing Steel in Concrete. In H. Lim (Ed.), *Handbook of Research on Recent Developments in Materials Science and Corrosion Engineering Education* (pp. 118–146). Hershey, PA: IGI Global. doi:10.4018/978-1-4666-8183-5.ch007

Jagadish,, & Bhowmik, S. (2019). Parameters Optimization of FDM for the Quality of Prototypes Using an Integrated MCDM Approach. In K. Kumar, D. Zindani, & J. Davim (Eds.), Additive Manufacturing Technologies From an Optimization Perspective (pp. 199-220). Hershey, PA: IGI Global. doi:10.4018/978-1-5225-9167-2.ch010

Jayalakshmi, S., & Singh, R. A. (2015). Processing Routes, Mechanical, and Tribological Properties of Light Metal Matrix Nanocomposites. In R. Tyagi & J. Davim (Eds.), *Processing Techniques and Tribological Behavior of Composite Materials* (pp. 1–46). Hershey, PA: IGI Global. doi:10.4018/978-1-4666-7530-8.ch001

Jayapalan, S. (2018). A Review of Chemical Treatments on Natural Fibers-Based Hybrid Composites for Engineering Applications. In K. Kumar & J. Davim (Eds.), *Composites and Advanced Materials for Industrial Applications* (pp. 16–37). Hershey, PA: IGI Global. doi:10.4018/978-1-5225-5216-1.ch002

Jucha, S. O., & Moskal, G. J. (2018). Thermal Barrier Coatings on the Base of Samarium Zirconates. In A. Pakseresht (Ed.), *Production, Properties, and Applications of High Temperature Coatings* (pp. 79–106). Hershey, PA: IGI Global. doi:10.4018/978-1-5225-4194-3.ch004

Kalita, H., Zindani, D., & Kumar, K. (2019). Additive Manufacturing: A Tool for Better Education. In K. Kumar, D. Zindani, & J. Davim (Eds.), *Additive Manufacturing Technologies From an Optimization Perspective* (pp. 41–76). Hershey, PA: IGI Global. doi:10.4018/978-1-5225-9167-2.ch003

Kamiya, H. (2014). Surface Modification and Structure Control for Nano- and Fine-Particle Aggregation and Adhesion Behaviour Control in Liquid Phase. In L. Santo & J. Davim (Eds.), *Surface Engineering Techniques and Applications: Research Advancements* (pp. 281–305). Hershey, PA: IGI Global. doi:10.4018/978-1-4666-5141-8.ch009

Kandasamy, J., & Sakthivel, A. R. K. E. K., V., M. S., S., M. T. H., S., Reddy, V. S., & Gutha, J. (2019). Assessment of Remanufacturability Index for an Automotive Product: A Case Study. In M. Uthayakumar, S. Raj, T. Ko, S. Kumaran, & J. Davim (Eds.), Handbook of Research on Green Engineering Techniques for Modern Manufacturing (pp. 296-308). Hershey, PA: IGI Global. doi:10.4018/978-1-5225-5445-5.ch016

Kandasamy, J., & Sakthivel, A. R. K. E. K., V., Reddy, V. S., & K. S., B. (2019). Application of Cluster Analysis for Identifying Potential Automotive Organizations Towards the Conduct of Green Manufacturing Sustainability Studies. In M. Uthayakumar, S. Raj, T. Ko, S. Kumaran, & J. Davim (Eds.), Handbook of Research on Green Engineering Techniques for Modern Manufacturing (pp. 309-322). Hershey, PA: IGI Global. doi:10.4018/978-1-5225-5445-5.ch017

Katrancioglu, S., Kilic, H. S., & Uslu, C. A. (2019). Solution Approaches for Reverse Logistics Considering Recovery Options: A Literature Review. In M. Uthayakumar, S. Raj, T. Ko, S. Kumaran, & J. Davim (Eds.), *Handbook of Research on Green Engineering Techniques for Modern Manufacturing* (pp. 192–210). Hershey, PA: IGI Global. doi:10.4018/978-1-5225-5445-5.ch011

Keshavamurthy, R., Naveena, B. E., & Sekhar, N. (2018). Thermal Spray Coatings for Erosion-Corrosion Protection. In A. Pakseresht (Ed.), *Production, Properties, and Applications of High Temperature Coatings* (pp. 246–267). Hershey, PA: IGI Global. doi:10.4018/978-1-5225-4194-3.ch010

Khan, L. A. (2015). Quickstep Processing of Polymeric Composites: An Out-Of-Autoclave (OOA) Approach. In R. Tyagi & J. Davim (Eds.), *Processing Techniques and Tribological Behavior of Composite Materials* (pp. 373–398). Hershey, PA: IGI Global. doi:10.4018/978-1-4666-7530-8.ch013

Kobayashi, T. (2017). Introduction of Environmental Materials. In T. Kobayashi (Ed.), *Applied Environmental Materials Science for Sustainability* (pp. 1–18). Hershey, PA: IGI Global. doi:10.4018/978-1-5225-1971-3.ch001

Kosaraju, S., Krishna Mohan, B., & Singh, S. K. (2019). Recent Advancement in Additive Manufacturing. In K. Kumar, D. Zindani, & J. Davim (Eds.), *Additive Manufacturing Technologies From an Optimization Perspective* (pp. 1–19). Hershey, PA: IGI Global. doi:10.4018/978-1-5225-9167-2.ch001

Kumar, K., & Davim, J. P. (2018). Comparative Study of Mechanical and Tribological Behaviour of Thermoplast-Based Composites: With Inorganic and Natural Particulate Fillers. In R. Somashekar, & T. Urs G. (Eds.), Emergent Research on Polymeric and Composite Materials (pp. 78-98). Hershey, PA: IGI Global. doi:10.4018/978-1-5225-3023-7.ch003

Kumar, M., Kumar, R., Kumar, S., & Prakash, C. (2019). Biomechanical Properties of Orthopedic and Dental Implants: A Comprehensive Review. In M. Uthayakumar, S. Raj, T. Ko, S. Kumaran, & J. Davim (Eds.), *Handbook of Research on Green Engineering Techniques for Modern Manufacturing* (pp. 1–13). Hershey, PA: IGI Global. doi:10.4018/978-1-5225-5445-5.ch001

Kumari, N., & Kumar, K. (2018). Efficacy of Composites for Fabrication of Orthotic Calipers. In R. Somashekar, & T. Urs G. (Eds.), Emergent Research on Polymeric and Composite Materials (pp. 267-286). Hershey, PA: IGI Global. doi:10.4018/978-1-5225-3023-7.ch012

Kumari, N., & Kumar, K. (2018). Fabrication of Orthotic Calipers With Epoxy-Based Green Composite. In K. Kumar & J. Davim (Eds.), *Composites and Advanced Materials for Industrial Applications* (pp. 157–176). Hershey, PA: IGI Global. doi:10.4018/978-1-5225-5216-1.ch008

Kuppusamy, R. R. (2018). Development of Aerospace Composite Structures Through Vacuum-Enhanced Resin Transfer Moulding Technology (VERTMTy): Vacuum-Enhanced Resin Transfer Moulding. In K. Kumar & J. Davim (Eds.), *Composites and Advanced Materials for Industrial Applications* (pp. 99–111). Hershey, PA: IGI Global. doi:10.4018/978-1-5225-5216-1.ch005

Kutschke, L. (2020). Social Life-Cycle Assessment for Building Materials. In B. Brownell (Ed.), *Examining the Environmental Impacts of Materials and Buildings* (pp. 217–250). Hershey, PA: IGI Global. doi:10.4018/978-1-7998-2426-8.ch008

Li, J., Xiong, D., Qin, Y., & Tyagi, R. (2015). Tribological Behavior of Ni-Based Self-Lubricating Composites at Elevated Temperatures. In R. Tyagi & J. Davim (Eds.), *Processing Techniques and Tribological Behavior of Composite Materials* (pp. 72–106). Hershey, PA: IGI Global. doi:10.4018/978-1-4666-7530-8.ch003

Li, K., & Kobayashi, T. (2017). Ionic Liquids and Poly (Ionic Liquid)s Used as Green Solvent and Ultrasound Responded Materials. In T. Kobayashi (Ed.), *Applied Environmental Materials Science for Sustainability* (pp. 327–346). Hershey, PA: IGI Global. doi:10.4018/978-1-5225-1971-3.ch015

Louhenkilpi, S., & Ganguly, S. (2016). Modeling of Steelmaking Processes. In S. Datta & J. Davim (Eds.), *Computational Approaches to Materials Design: Theoretical and Practical Aspects* (pp. 369–421). Hershey, PA: IGI Global. doi:10.4018/978-1-5225-0290-6.ch013

Lourenço, E. J., Moita, N., Esteves, S., Peças, P., Ribeiro, I. I., Pereira, J. P., & Oliveira, L. M. (2019). Multi-Perspective Eco-Efficiency Assessment to Foster Sustainability in Plastic Parts Production: An Integrated Tool for Industrial Use. In M. Uthayakumar, S. Raj, T. Ko, S. Kumaran, & J. Davim (Eds.), *Handbook of Research on Green Engineering Techniques for Modern Manufacturing* (pp. 212–249). Hershey, PA: IGI Global. doi:10.4018/978-1-5225-5445-5.ch012

Loy, J., & Schork, T. (2019). Building Relationships: Changing Technology and Society. In G. Koç & B. Christiansen (Eds.), *Reusable and Sustainable Building Materials in Modern Architecture* (pp. 166–187). Hershey, PA: IGI Global. doi:10.4018/978-1-5225-6995-4.ch008

M., S. M., & V., A. (2019). Kinematic Modelling and Simulation of 8 Degrees of Freedom SCARA Robot. In M. Uthayakumar, S. Raj, T. Ko, S. Kumaran, & J. Davim (Eds.), Handbook of Research on Green Engineering Techniques for Modern Manufacturing (pp. 77-97). Hershey, PA: IGI Global. doi:10.4018/978-1-5225-5445-5.ch005

Ma, W., Shang, J., Lu, J., & Meng, J. (2015). Material and Tribology Issues of Self-Lubricating Copper Matrix Composite. In R. Tyagi & J. Davim (Eds.), *Processing Techniques and Tribological Behavior of Composite Materials* (pp. 47–71). Hershey, PA: IGI Global. doi:10.4018/978-1-4666-7530-8.ch002

Mabbett, I. (2015). Applying a Coherent Academy Training Structure to Vertically Integrate Learning, Teaching and Personal Development in Materials Science and Engineering. In H. Lim (Ed.), *Handbook of Research on Recent Developments in Materials Science and Corrosion Engineering Education* (pp. 82–99). Hershey, PA: IGI Global. doi:10.4018/978-1-4666-8183-5.ch005

Mahamood, R. M. (2016). Laser Metal Deposition Process. In E. Akinlabi, R. Mahamood, & S. Akinlabi (Eds.), *Advanced Manufacturing Techniques Using Laser Material Processing* (pp. 46–59). Hershey, PA: IGI Global. doi:10.4018/978-1-5225-0329-3.ch003

Mahamood, R. M., & Akinlabi, E. T. (2016). Laser Additive Manufacturing. In E. Akinlabi, R. Mahamood, & S. Akinlabi (Eds.), *Advanced Manufacturing Techniques Using Laser Material Processing* (pp. 1–23). Hershey, PA: IGI Global. doi:10.4018/978-1-5225-0329-3.ch001

Mahamood, R. M., Akinlabi, E. T., Shukla, M., & Pityana, S. (2014). Improving Surface Integrity Using Laser Metal Deposition Process. In L. Santo & J. Davim (Eds.), *Surface Engineering Techniques and Applications: Research Advancements* (pp. 146–176). Hershey, PA: IGI Global. doi:10.4018/978-1-4666-5141-8.ch005

Mahamood, R. M., Shukla, M., & Pityana, S. (2014). Laser Additive Manufacturing in Surface Modification of Metals. In L. Santo & J. Davim (Eds.), *Surface Engineering Techniques and Applications: Research Advancements* (pp. 222–248). Hershey, PA: IGI Global. doi:10.4018/978-1-4666-5141-8.ch007

Mahmood, W. A., & Azarian, M. H. (2017). Inorganic-Organic Composite Materials from Liquid Natural Rubber and Epoxidised Natural Rubber Derivatives: Prospects and Applications. In T. Kobayashi (Ed.), *Applied Environmental Materials Science for Sustainability* (pp. 128–140). Hershey, PA: IGI Global. doi:10.4018/978-1-5225-1971-3.ch006

Maina, M. R. (2016). Laser Additive Manufacturing of Titanium-Based Implants: A Review. In E. Akinlabi, R. Mahamood, & S. Akinlabi (Eds.), *Advanced Manufacturing Techniques Using Laser Material Processing* (pp. 236–247). Hershey, PA: IGI Global. doi:10.4018/978-1-5225-0329-3.ch009

Mainali, B., Petrolito, J., Russell, J., Ionescu, D., & Al Abadi, H. (2015). Integrating Sustainable Engineering Principles in Material Science Engineering Education. In H. Lim (Ed.), *Handbook of Research on Recent Developments in Materials Science and Corrosion Engineering Education* (pp. 273–291). Hershey, PA: IGI Global. doi:10.4018/978-1-4666-8183-5.ch014

Majumdar, J. D., Weisheit, A., & Manna, I. (2016). Laser Surface Processing for Tailoring of Properties by Optimization of Microstructure. In E. Akinlabi, R. Mahamood, & S. Akinlabi (Eds.), *Advanced Manufacturing Techniques Using Laser Material Processing* (pp. 121–171). Hershey, PA: IGI Global. doi:10.4018/978-1-5225-0329-3.ch006

Malik, F. M., Ali, S. F., Bal, B., & Kececi, E. F. (2019). Determination of Optimum Process Parameter Values in Additive Manufacturing for Impact Resistance. In K. Kumar, D. Zindani, & J. Davim (Eds.), *Additive Manufacturing Technologies From an Optimization Perspective* (pp. 221–234). Hershey, PA: IGI Global. doi:10.4018/978-1-5225-9167-2.ch011

Mans, J. W. (2020). Post-Normal Material Practice: "Building" as Inquiry. In B. Brownell (Ed.), *Examining the Environmental Impacts of Materials and Buildings* (pp. 282–312). Hershey, PA: IGI Global. doi:10.4018/978-1-7998-2426-8.ch010

Marazani, T., Madyira, D. M., & Akinlabi, E. T. (2019). Additive Manufacturing for Crack Repair Applications in Metals: A Case of Titanium (Ti) Alloys. In K. Kumar, D. Zindani, & J. Davim (Eds.), *Additive Manufacturing Technologies From an Optimization Perspective* (pp. 77–101). Hershey, PA: IGI Global. doi:10.4018/978-1-5225-9167-2.ch004

Mirza, J., Hussin, M. W., & Ismail, M. A. (2019). Characterization and Assessment of Iron Ore Tailings as Raw Materials for Construction Industries. In *Recycled Waste Materials in Concrete Construction: Emerging Research and Opportunities* (pp. 1–16). Hershey, PA: IGI Global. doi:10.4018/978-1-5225-8325-7.ch001

Mirza, J., Hussin, M. W., & Ismail, M. A. (2019). Comparison Between Ordinary Portland Cement and Geopolymer Concretes Against Sulphuric Acid Attack. In *Recycled Waste Materials in Concrete Construction: Emerging Research and Opportunities* (pp. 126–134). Hershey, PA: IGI Global. doi:10.4018/978-1-5225-8325-7.ch009

Mirza, J., Hussin, M. W., & Ismail, M. A. (2019). Durability of Mortars Containing Ceramic Tile Waste Exposed to Sulphate Attack. In *Recycled Waste Materials in Concrete Construction: Emerging Research and Opportunities* (pp. 86–98). Hershey, PA: IGI Global. doi:10.4018/978-1-5225-8325-7.ch006

Mirza, J., Hussin, M. W., & Ismail, M. A. (2019). Effect of High-Volume Oil Palm Biomass Waste in Mortar. In *Recycled Waste Materials in Concrete Construction: Emerging Research and Opportunities* (pp. 17–30). Hershey, PA: IGI Global. doi:10.4018/978-1-5225-8325-7.ch002

Mirza, J., Hussin, M. W., & Ismail, M. A. (2019). Effect of Metakaolin on Geopolymer Industry. In *Recycled Waste Materials in Concrete Construction: Emerging Research and Opportunities* (pp. 49–70). Hershey, PA: IGI Global. doi:10.4018/978-1-5225-8325-7.ch004

Mirza, J., Hussin, M. W., & Ismail, M. A. (2019). Effects of Palm Oil Fuel Ash as Micro-Filler on Interfacial Porosity of Polymer Concrete. In *Recycled Waste Materials in Concrete Construction: Emerging Research and Opportunities* (pp. 99–111). Hershey, PA: IGI Global. doi:10.4018/978-1-5225-8325-7.ch007

Mirza, J., Hussin, M. W., & Ismail, M. A. (2019). Evaluation of Electric Arc Furnace Oxidizing Slag Aggregates Quality and Development of Functional Concrete. In *Recycled Waste Materials in Concrete Construction: Emerging Research and Opportunities* (pp. 112–125). Hershey, PA: IGI Global. doi:10.4018/978-1-5225-8325-7.ch008

Mirza, J., Hussin, M. W., & Ismail, M. A. (2019). Performance of Mortar Grouts Incorporating Rice Husk Ash and Fly Ash. In *Recycled Waste Materials in Concrete Construction: Emerging Research and Opportunities* (pp. 31–48). Hershey, PA: IGI Global. doi:10.4018/978-1-5225-8325-7.ch003

Mirza, J., Hussin, M. W., & Ismail, M. A. (2019). Self-Healing Mortar. In *Recycled Waste Materials in Concrete Construction: Emerging Research and Opportunities* (pp. 135–148). Hershey, PA: IGI Global. doi:10.4018/978-1-5225-8325-7.ch010

Mirza, J., Hussin, M. W., & Ismail, M. A. (2019). Versatility of Cockle Shell in Concrete: A Conspectus. In *Recycled Waste Materials in Concrete Construction: Emerging Research and Opportunities* (pp. 71–85). Hershey, PA: IGI Global. doi:10.4018/978-1-5225-8325-7.ch005

Mohan, S., Gautam, R. K., & Mohan, A. (2015). Tribology and Aluminium Matrix Composites. In R. Tyagi & J. Davim (Eds.), *Processing Techniques and Tribological Behavior of Composite Materials* (pp. 126–148). Hershey, PA: IGI Global. doi:10.4018/978-1-4666-7530-8.ch005

Mohanty, I., & Bhattacherjee, D. (2016). Artificial Neural Network and Its Application in Steel Industry. In S. Datta & J. Davim (Eds.), *Computational Approaches to Materials Design: Theoretical and Practical Aspects* (pp. 267–300). Hershey, PA: IGI Global. doi:10.4018/978-1-5225-0290-6.ch010

More, S., & Sutaria, M. (2020). Green Charcoal: Developing Biodegradable Construction Materials for a Circular Economy. In B. Brownell (Ed.), *Examining the Environmental Impacts of Materials and Buildings* (pp. 132–159). Hershey, PA: IGI Global. doi:10.4018/978-1-7998-2426-8.ch005

Mukherjee, A., & Mitra, P. (2018). Physical Characteristics of CBD Synthesized CdZnS Thin Films: CdZnS Thin Film Preparation Techniques, Properties, and Characterization. In R. Somashekar, & T. Urs G. (Eds.), Emergent Research on Polymeric and Composite Materials (pp. 186-208). Hershey, PA: IGI Global. doi:10.4018/978-1-5225-3023-7.ch008

Mukhopadhyay, A., Barman, T. K., & Sahoo, P. (2018). Electroless Nickel Coatings for High Temperature Applications. In K. Kumar & J. Davim (Eds.), *Composites and Advanced Materials for Industrial Applications* (pp. 297–331). Hershey, PA: IGI Global. doi:10.4018/978-1-5225-5216-1.ch013

Nandi, A. K., & Deb, K. (2016). Multi-Objective Evolutionary Algorithms: Application in Designing Particle Reinforced Mould Materials. In S. Datta & J. Davim (Eds.), *Computational Approaches to Materials Design: Theoretical and Practical Aspects* (pp. 301–345). Hershey, PA: IGI Global. doi:10.4018/978-1-5225-0290-6.ch011

Novak, J. I., Liu, M. Z., & Loy, J. (2019). Designing Thin 2.5D Parts Optimized for Fused Deposition Modeling. In K. Kumar, D. Zindani, & J. Davim (Eds.), *Additive Manufacturing Technologies From an Optimization Perspective* (pp. 134–164). Hershey, PA: IGI Global. doi:10.4018/978-1-5225-9167-2.ch007

O'Donnell, C., & Pranger, D. (2019). Rethinking Waste Through Design. In G. Koç & B. Christiansen (Eds.), *Reusable and Sustainable Building Materials in Modern Architecture* (pp. 93–107). Hershey, PA: IGI Global. doi:10.4018/978-1-5225-6995-4.ch005

Odetola, P. I., Popoola, P. A., & Oladijo, P. (2018). Thin Coating Deposition by Magnetron Sputtering. In A. Pakseresht (Ed.), *Production, Properties, and Applications of High Temperature Coatings* (pp. 403–428). Hershey, PA: IGI Global. doi:10.4018/978-1-5225-4194-3.ch015

Oliveira, M. R., Jorge, D., & Peças, P. (2019). Methodology of Operationalization of KPIs for Shop-Floor. In M. Uthayakumar, S. Raj, T. Ko, S. Kumaran, & J. Davim (Eds.), *Handbook of Research on Green Engineering Techniques for Modern Manufacturing* (pp. 163–191). Hershey, PA: IGI Global. doi:10.4018/978-1-5225-5445-5.ch010

Onutai, S., Jiemsirilers, S., & Kobayashi, T. (2017). Geopolymer Sourced with Fly Ash and Industrial Aluminum Waste for Sustainable Materials. In T. Kobayashi (Ed.), *Applied Environmental Materials Science for Sustainability* (pp. 165–185). Hershey, PA: IGI Global. doi:10.4018/978-1-5225-1971-3.ch008

Osorno, D. M., & Castro, C. (2018). Cellulose Application in Food Industry: A Review. In R. Somashekar, & T. Urs G. (Eds.), Emergent Research on Polymeric and Composite Materials (pp. 38-77). Hershey, PA: IGI Global. doi:10.4018/978-1-5225-3023-7.ch002

Ozturk, F., Deveci, T., Gunister, E., & Simmons, R. J. (2015). Innovative Instructional Strategies for Teaching Materials Science in Engineering. In H. Lim (Ed.), *Handbook of Research on Recent Developments in Materials Science and Corrosion Engineering Education* (pp. 100–117). Hershey, PA: IGI Global. doi:10.4018/978-1-4666-8183-5.ch006

Parikh, H. H., & Gohil, P. P. (2015). Composites as TRIBO Materials in Engineering Systems: Significance and Applications. In R. Tyagi & J. Davim (Eds.), *Processing Techniques and Tribological Behavior of Composite Materials* (pp. 168–191). Hershey, PA: IGI Global. doi:10.4018/978-1-4666-7530-8.ch007

Paul, S. R., Nayak, S. K., Anis, A., & Pal, K. (2018). Methods of Preparation of Nanoparticle Formulations for the Treatment of COPD: A Short Review. In R. Somashekar, & T. Urs G. (Eds.), Emergent Research on Polymeric and Composite Materials (pp. 247-266). Hershey, PA: IGI Global. doi:10.4018/978-1-5225-3023-7.ch011

Persaud-Sharma, D. (2014). Surface Engineering Techniques and Applications: Research Advancements. In L. Santo & J. Davim (Eds.), *Surface Engineering Techniques and Applications: Research Advancements* (pp. 73–101). Hershey, PA: IGI Global. doi:10.4018/978-1-4666-5141-8.ch003

Potiyaraj, P. (2017). Poly (Lactic Acid) Generated for Advanced Materials. In T. Kobayashi (Ed.), *Applied Environmental Materials Science for Sustainability* (pp. 106–127). Hershey, PA: IGI Global. doi:10.4018/978-1-5225-1971-3.ch005

Prakash, C., Singh, S., Abdul-Rani, A. M., Uddin, M. S., Pabla, B. S., & Puri, S. (2019). Spark Plasma Sintering of Mg-Zn-Mn-Si-HA Alloy for Bone Fixation Devices: Fabrication of Biodegradable Low Elastic Porous Mg-Zn-Mn-Si-HA Alloy. In M. Uthayakumar, S. Raj, T. Ko, S. Kumaran, & J. Davim (Eds.), *Handbook of Research on Green Engineering Techniques for Modern Manufacturing* (pp. 282–295). Hershey, PA: IGI Global. doi:10.4018/978-1-5225-5445-5.ch015

Puglisi, V., & Ciaramella, A. (2019). Urban Quality Assessment at the Neighborhood Scale: An Experimental Approach. In G. Koç & B. Christiansen (Eds.), *Reusable and Sustainable Building Materials in Modern Architecture* (pp. 188–220). Hershey, PA: IGI Global. doi:10.4018/978-1-5225-6995-4.ch009

Purohit, P., & Vagge, S. T. (2018). Oxidation Behavior of LTA/YSZ Intermixed Layer Coating. In A. Pakseresht (Ed.), *Production, Properties, and Applications of High Temperature Coatings* (pp. 107–130). Hershey, PA: IGI Global. doi:10.4018/978-1-5225-4194-3.ch005

R., K., Tambrallimath, V., Kuppahalli, P., & N., S. (2019). Additive Manufacturing Process and Their Applications for Green Technology. In M. Uthayakumar, S. Raj, T. Ko, S. Kumaran, & J. Davim (Eds.), Handbook of Research on Green Engineering Techniques for Modern Manufacturing (pp. 262-281). Hershey, PA: IGI Global. doi:10.4018/978-1-5225-5445-5.ch014

Rajkumar, K., & Aravindan, S. (2015). Tribological Characteristics of Copper-Nano Carbon Crystalline Composites. In R. Tyagi & J. Davim (Eds.), *Processing Techniques and Tribological Behavior of Composite Materials* (pp. 107–125). Hershey, PA: IGI Global. doi:10.4018/978-1-4666-7530-8.ch004

Ramdani, N., & Azibi, M. (2018). Polymer Composite Materials for Microelectronics Packaging Applications: Composites for Microelectronics Packaging. In K. Kumar & J. Davim (Eds.), *Composites and Advanced Materials for Industrial Applications* (pp. 177–211). Hershey, PA: IGI Global. doi:10.4018/978-1-5225-5216-1.ch009

Ramdani, N., & Azibi, M. (2018). Properties Enhancement of Polymer/Ceramic Nanocomposites: Polymer Nanocomposites. In R. Somashekar, & T. Urs G. (Eds.), Emergent Research on Polymeric and Composite Materials (pp. 99-121). Hershey, PA: IGI Global. doi:10.4018/978-1-5225-3023-7.ch004

Ripple, J. (2020). Building Codes Don't Measure Up: A Case for Urban Material Performance Standards. In B. Brownell (Ed.), *Examining the Environmental Impacts of Materials and Buildings* (pp. 32–63). Hershey, PA: IGI Global. doi:10.4018/978-1-7998-2426-8.ch002

S., R., & Siddaramaiah,. (2018). Effect of Cenosphere on the Performance of Polyurethane/Polystyrene Interpenetrating Polymer Network Green Composites. In R. Somashekar, & T. Urs G. (Eds.), Emergent Research on Polymeric and Composite Materials (pp. 158-185). Hershey, PA: IGI Global. doi:10.4018/978-1-5225-3023-7.ch007

Sadri, E., & Ashrafizadeh, F. (2018). High Temperature Nanocomposite Coatings by Plasma Spraying for Friction and Wear Applications. In A. Pakseresht (Ed.), *Production, Properties, and Applications of High Temperature Coatings* (pp. 216–245). Hershey, PA: IGI Global. doi:10.4018/978-1-5225-4194-3.ch009

Sagiri, S. S., Nayak, S. K., Lakshmi, S., & Pal, K. (2018). Preparation and Characterization of Biopolymeric Nanoparticles as Drug Delivery Vehicles: A Comparative Study. In R. Somashekar, & T. Urs G. (Eds.), Emergent Research on Polymeric and Composite Materials (pp. 225-246). Hershey, PA: IGI Global. doi:10.4018/978-1-5225-3023-7.ch010

Sahoo, P., Roy, S., & Davim, J. P. (2014). Design and Selection of Chemically Deposited Ni-P-W Coatings for Optimum Tribological Behavior. In L. Santo & J. Davim (Eds.), *Surface Engineering Techniques and Applications: Research Advancements* (pp. 45–72). Hershey, PA: IGI Global. doi:10.4018/978-1-4666-5141-8.ch002

Sahoo, S. (2018). Laminated Composite Hypar Shells as Roofing Units: Static and Dynamic Behavior. In K. Kumar & J. Davim (Eds.), *Composites and Advanced Materials for Industrial Applications* (pp. 249–269). Hershey, PA: IGI Global. doi:10.4018/978-1-5225-5216-1.ch011

Sakulich, A., El-Korchi, T., & Sisson, R. D. Jr. (2015). The Interdisciplinary, Project-Based Infrastructure Degradation Curriculum at Worcester Polytechnic Institute. In H. Lim (Ed.), *Handbook of Research on Recent Developments in Materials Science and Corrosion Engineering Education* (pp. 67–81). Hershey, PA: IGI Global. doi:10.4018/978-1-4666-8183-5.ch004

Satoh, M. (2017). Metal Ion Separation with Functional Adsorbents and Phytoremediation Used as Sustainable Technologies. In T. Kobayashi (Ed.), *Applied Environmental Materials Science for Sustainability* (pp. 284–312). Hershey, PA: IGI Global. doi:10.4018/978-1-5225-1971-3.ch013

Sezgin, H., & Berkalp, O. B. (2018). Textile-Reinforced Composites for the Automotive Industry. In K. Kumar & J. Davim (Eds.), *Composites and Advanced Materials for Industrial Applications* (pp. 129–156). Hershey, PA: IGI Global. doi:10.4018/978-1-5225-5216-1.ch007

Shahien, M. (2018). Reactive Plasma Spray: A Method for Nitride Coatings Deposition in Thermal Spray. In A. Pakseresht (Ed.), *Production, Properties, and Applications of High Temperature Coatings* (pp. 299–332). Hershey, PA: IGI Global. doi:10.4018/978-1-5225-4194-3.ch012

Sharma, N., & Kumar, K. (2018). Fabrication of Porous NiTi Alloy Using Organic Binders. In K. Kumar & J. Davim (Eds.), *Composites and Advanced Materials for Industrial Applications* (pp. 38–62). Hershey, PA: IGI Global. doi:10.4018/978-1-5225-5216-1.ch003

Shukla, R., Anapagaddi, R., Singh, A. K., Allen, J. K., Panchal, J. H., & Mistree, F. (2016). Integrated Computational Materials Engineering for Determining the Set Points of Unit Operations for Production of a Steel Product Mix. In S. Datta & J. Davim (Eds.), *Computational Approaches to Materials Design: Theoretical and Practical Aspects* (pp. 163–191). Hershey, PA: IGI Global. doi:10.4018/978-1-5225-0290-6.ch006

Singh, I., & Debnath, K. (2015). Advanced Machining Techniques for Fiber-Reinforced Polymer Composites. In R. Tyagi & J. Davim (Eds.), *Processing Techniques and Tribological Behavior of Composite Materials* (pp. 317–340). Hershey, PA: IGI Global. doi:10.4018/978-1-4666-7530-8.ch011

Singh, R., & Dutta, S. (2018). Visible Light Active Nanocomposites for Photocatalytic Applications. In K. Kumar & J. Davim (Eds.), *Composites and Advanced Materials for Industrial Applications* (pp. 270–296). Hershey, PA: IGI Global. doi:10.4018/978-1-5225-5216-1.ch012

Singh, S., Prakash, C., & Uthayakumar, M. (2019). Recent Advancements in Customized Investment Castings Through Additive Manufacturing: Implication of Additive Manufacturing in Investment Casting. In M. Uthayakumar, S. Raj, T. Ko, S. Kumaran, & J. Davim (Eds.), *Handbook of Research on Green Engineering Techniques for Modern Manufacturing* (pp. 24–48). Hershey, PA: IGI Global. doi:10.4018/978-1-5225-5445-5.ch003

Stavropoulos, P., Koutsomichalis, A., & Vaxevanidis, N. (2016). Laser-Based Manufacturing Processes for Aerospace Applications. In E. Akinlabi, R. Mahamood, & S. Akinlabi (Eds.), *Advanced Manufacturing Techniques Using Laser Material Processing* (pp. 24–45). Hershey, PA: IGI Global. doi:10.4018/978-1-5225-0329-3.ch002

Subramani, N. K. K. N., S., Shivanna, S., B. M., J. R., & Hatna, S. (2018). Advances in Polymer-Based Piezoelectric Systems. In R. Somashekar, & T. Urs G. (Eds.), Emergent Research on Polymeric and Composite Materials (pp. 139-157). Hershey, PA: IGI Global. doi:10.4018/978-1-5225-3023-7.ch006

Surekha, K., & Akinlabi, E. (2014). Surface Engineering by Friction Stir Processing and Friction Surfacing. In L. Santo & J. Davim (Eds.), *Surface Engineering Techniques and Applications: Research Advancements* (pp. 102–145). Hershey, PA: IGI Global. doi:10.4018/978-1-4666-5141-8.ch004

Takahashi, Y. (2017). Eco-Friendly On-Site Water Analyses for Ultra-Trace Harmful Ions. In T. Kobayashi (Ed.), *Applied Environmental Materials Science for Sustainability* (pp. 313–326). Hershey, PA: IGI Global. doi:10.4018/978-1-5225-1971-3.ch014

Taokaew, S., Phisalaphong, M., & Newby, B. Z. (2017). Bacterial Cellulose: Biosyntheses, Modifications, and Applications. In T. Kobayashi (Ed.), *Applied Environmental Materials Science for Sustainability* (pp. 255–283). Hershey, PA: IGI Global. doi:10.4018/978-1-5225-1971-3.ch012

Thejas Urs, G., & Somashekar, R. (2018). Fibre Diffraction and Whole Powder Pattern Fitting in Polymers. In R. Somashekar, & T. Urs G. (Eds.), Emergent Research on Polymeric and Composite Materials (pp. 1-37). Hershey, PA: IGI Global. doi:10.4018/978-1-5225-3023-7.ch001

Tlotleng, M., Akinlabi, E. T., Shukla, M., & Pityana, S. (2014). Application of Laser Assisted Cold Spraying Process for Materials Deposition. In L. Santo & J. Davim (Eds.), *Surface Engineering Techniques and Applications: Research Advancements* (pp. 177–221). Hershey, PA: IGI Global. doi:10.4018/978-1-4666-5141-8.ch006

Toptan, F., & Rocha, L. A. (2015). Tribocorrosion in Metal Matrix Composites. In R. Tyagi & J. Davim (Eds.), *Processing Techniques and Tribological Behavior of Composite Materials* (pp. 149–167). Hershey, PA: IGI Global. doi:10.4018/978-1-4666-7530-8.ch006

Trang, T. T. (2017). Study on the Use of Biomass Polymer Sheets in Water/Alcohol Pervaporation as a Sustainable Source of Alcohol Energy. In T. Kobayashi (Ed.), *Applied Environmental Materials Science for Sustainability* (pp. 234–254). Hershey, PA: IGI Global. doi:10.4018/978-1-5225-1971-3.ch011

Trubiano, F. (2020). Toxicity in Architectural Plastics: Life-Cycle Index of Human Health in Building (LCI-HHB). In B. Brownell (Ed.), *Examining the Environmental Impacts of Materials and Buildings* (pp. 103–131). Hershey, PA: IGI Global. doi:10.4018/978-1-7998-2426-8.ch004

Umm-i-Kalsoom. Ali, N., Begum, N., Hussnain, A., & Ahmad, R. (2018). Hardness of Duplex Treated Stainless Steel: Influence of the Architecture and Composition of Films. In A. Pakseresht (Ed.), Production, Properties, and Applications of High Temperature Coatings (pp. 447-466). Hershey, PA: IGI Global. doi:10.4018/978-1-5225-4194-3.ch017

Uthayakumar, M., Karnan, B., Slota, A., Zajac, J., & Davim, J. P. (2019). Performance Study of LaPO4-Y2O3 Composite Fabricated by Sol-Gel Process Using Abrasive Waterjet Machining. In M. Uthayakumar, S. Raj, T. Ko, S. Kumaran, & J. Davim (Eds.), *Handbook of Research on Green Engineering Techniques for Modern Manufacturing* (pp. 143–161). Hershey, PA: IGI Global. doi:10.4018/978-1-5225-5445-5.ch009

Vergara, D., Lorenzo, M., & Rubio, M. (2015). Virtual Environments in Materials Science and Engineering: The Students' Opinion. In H. Lim (Ed.), *Handbook of Research on Recent Developments in Materials Science and Corrosion Engineering Education* (pp. 148–165). Hershey, PA: IGI Global. doi:10.4018/978-1-4666-8183-5.ch008

Verma, R., Suri, N. M., & Kant, S. (2018). Slurry Sprayed Mullite Coatings and Their Corrosion Performances. In A. Pakseresht (Ed.), *Production, Properties, and Applications of High Temperature Coatings* (pp. 187–214). Hershey, PA: IGI Global. doi:10.4018/978-1-5225-4194-3.ch008

Waghmode, M. S., Gunjal, A. B., Bhujbal, N. N., Patil, N. N., & Nawani, N. N. (2019). Eco-Friendly Construction. In G. Koç & B. Christiansen (Eds.), *Reusable and Sustainable Building Materials in Modern Architecture* (pp. 80–92). Hershey, PA: IGI Global. doi:10.4018/978-1-5225-6995-4.ch004

Wesley, D. T., & Puffer, S. M. (2019). The End of Sand: Confronting One of the Greatest Environmental Challenges of the New Millennium. In G. Koç & B. Christiansen (Eds.), *Reusable and Sustainable Building Materials in Modern Architecture* (pp. 1–27). Hershey, PA: IGI Global. doi:10.4018/978-1-5225-6995-4.ch001

Xie, W. (2018). Thermal Spray Methods and Splat Formation. In A. Pakseresht (Ed.), *Production, Properties, and Applications of High Temperature Coatings* (pp. 1–24). Hershey, PA: IGI Global. doi:10.4018/978-1-5225-4194-3.ch001

Yagoube, I. I., Ibrahim, T., Nor, N. M., & Nallagownden, P. (2019). Performance of PM Linear Generator Under Various Ferromagnetic Materials for Wave Energy Conversion. In M. Uthayakumar, S. Raj, T. Ko, S. Kumaran, & J. Davim (Eds.), *Handbook of Research on Green Engineering Techniques for Modern Manufacturing* (pp. 113–126). Hershey, PA: IGI Global. doi:10.4018/978-1-5225-5445-5.ch007

Zhu, S., & Wang, F. (2018). Nanocrystalline, Enamel and Composite Coatings for Superalloys. In A. Pakseresht (Ed.), *Production, Properties, and Applications of High Temperature Coatings* (pp. 160–186). Hershey, PA: IGI Global. doi:10.4018/978-1-5225-4194-3.ch007

Zindani, D., Faisal, N., & Kumar, K. (2019). Optimization of Process Parameters for Electro-Chemical Machining of EN19: Using Particle Swamp Optimization. In M. Uthayakumar, S. Raj, T. Ko, S. Kumaran, & J. Davim (Eds.), *Handbook of Research on Green Engineering Techniques for Modern Manufacturing* (pp. 127–142). Hershey, PA: IGI Global. doi:10.4018/978-1-5225-5445-5.ch008

Zindani, D., & Kumar, K. (2018). Industrial Applications of Polymer Composite Materials. In K. Kumar & J. Davim (Eds.), *Composites and Advanced Materials for Industrial Applications* (pp. 1–15). Hershey, PA: IGI Global. doi:10.4018/978-1-5225-5216-1.ch001

Zindani, D., & Kumar, K. (2019). Integrated Manufacturing System for Complex Geometries: Towards Zero Waste in Additive Manufacturing. In M. Uthayakumar, S. Raj, T. Ko, S. Kumaran, & J. Davim (Eds.), *Handbook of Research on Green Engineering Techniques for Modern Manufacturing* (pp. 14–23). Hershey, PA: IGI Global. doi:10.4018/978-1-5225-5445-5.ch002

Zindani, D., Maity, S. R., & Bhowmik, S. (2018). A Decision-Making Approach for Material Selection of Polymeric Composite Bumper Beam. In K. Kumar & J. Davim (Eds.), *Composites and Advanced Materials for Industrial Applications* (pp. 112–128). Hershey, PA: IGI Global. doi:10.4018/978-1-5225-5216-1.ch006

About the Author

Barbara Cięcińska is the lecturer and researcher at Rzeszow University of Technology (Poland) since 1997. She is interested in modern and innovative technologies in machine building. For many years she has been conducting research on the possibilities of application of unconventional methods in order to minimize materials use and the development of parameters of machine parts. The research was crowned in 2004 by a PhD thesis on shot peening technology and its application for the improvement of fatigue strength in unconventional parts of the machine (parts with decarburized surface layer). Currently her main area of interest are the problems of restoring exploitation properties to machine parts, their quality and durability. For several years, she has been conducting experimental research in the field of adhesive bonding, paying particular attention to the chemicals used in this process and their harmfulness to human and the environment. She carries out research on the possibility of replacing harmful processes of surface preparation before adhesive bonding and improving the adhesion conditions by laser beam treatment. In addition, in the last two years, she has extended the scope of research to include laser cutting technology for plastics, as one of non-waste methods in machine building. She has cooperated with plants and research-educational organization in the laser treatment usage area for five years. She conducts her research closely with the Technical University in Kosice, the Faculty of Mathematics and Natural Sciences at University of Rzeszow, the Faculty of Mathematics and Applied Physics at Rzeszow University of Technology, TRUMPF Poland and SPI Lasers United Kingdom. She is a member of the Polish Society of Production Management, the Science Council of Technical Issues Journal and the Science for Industry and the Environment Foundation.

Index

P

Q

R

S

T

W

www.ingramcontent.com/pod-product-compliance
Lightning Source LLC
LaVergne TN
LVHW081324300926
849161LV00008B/1871